T0344595

# A Student's Guide to Special Relativity

This compact yet informative Guide presents an accessible route through Special Relativity, taking a modern axiomatic and geometrical approach. It begins by explaining key concepts and introducing Einstein's postulates. The consequences of the postulates – length contraction and time dilation – are unravelled qualitatively and then quantitatively. These strands are then tied together using the mathematical framework of the Lorentz transformation, before applying these ideas to kinematics and dynamics. This volume demonstrates the essential simplicity of the core ideas of Special Relativity, while acknowledging the challenges of developing new intuitions and dealing with the apparent paradoxes that arise. A valuable supplementary resource for intermediate undergraduates, as well as independent learners with some technical background, the Guide includes numerous exercises with hints and notes provided online. It lays the foundations for further study in General Relativity, which is introduced briefly in an appendix.

NORMAN GRAY is in the School of Physics and Astronomy, University of Glasgow, where he has, over the last 20 years, taught courses covering both special and general relativity. Since his PhD in particle theory, he has had a varied research career, most recently concentrating on the interface between astronomy and computing. He is author of *A Student's Guide to General Relativity* (2019), which has been a successful addition to Cambridge University Press's growing series of Student's Guides.

**Other books in the Student's Guide series:**

# A Student's Guide to Special Relativity

NORMAN GRAY
*University of Glasgow*

CAMBRIDGE
UNIVERSITY PRESS

# CAMBRIDGE
## UNIVERSITY PRESS

University Printing House, Cambridge CB2 8BS, United Kingdom

One Liberty Plaza, 20th Floor, New York, NY 10006, USA

477 Williamstown Road, Port Melbourne, VIC 3207, Australia

314–321, 3rd Floor, Plot 3, Splendor Forum, Jasola District Centre, New Delhi – 110025, India

103 Penang Road, #05–06/07, Visioncrest Commercial, Singapore 238467

Cambridge University Press is part of the University of Cambridge.

It furthers the University's mission by disseminating knowledge in the pursuit of education, learning, and research at the highest international levels of excellence.

www.cambridge.org
Information on this title: www.cambridge.org/highereducation/isbn/9781108834094
DOI: 10.1017/9781108999588

First published 2022

A catalogue record for this publication is available from the British Library.

ISBN 978-1-108-83409-4 Hardback
ISBN 978-1-108-99563-4 Paperback

Additional resources for this publication at cambridge.org/gray-sgsr

I do not know what I may appear to the world, but to myself I seem to have been only like a boy playing on the sea-shore, and diverting myself in now and then finding a smoother pebble or a prettier shell than ordinary, whilst the great ocean of truth lay all undiscovered before me.

Isaac Newton, as quoted in Brewster, *Memoirs of the Life, Writings, and Discoveries of Sir Isaac Newton*, Vol. 2, p. 407.

The eternally incomprehensible thing about the universe is its comprehensibility.

Albert Einstein (1936*a*)

# Contents

# Preface

This book is an introduction to Special Relativity (SR), which intends to be both comprehensive, in the sense of covering most of what a physicist will be presumed to know about relativity, and accessible to those at a reasonably early stage of their physics education. These aims are not incompatible. Learning about relativity does not require a huge volume of material to be learned, and the maths is not elaborate (the most advanced maths we use is basic calculus).

The very significant challenge to approaching this area of physics is that it is probably the first area we encounter where we cannot rely on our intuition, and where we must rely on abstract arguments to navigate past our intuitions to somewhere we could not otherwise reach. I fully embrace this in the axiomatic approach of this book: we start with the two axioms or postulates of SR, and then follow their consequences.

But SR is not *just* an abstraction. It is instead a foundational theory of how our physical world is constructed, and the consequences we deduce must match the physical world we find ourselves in. Making the link from abstraction to nature is of course a little tricky – there are a lot of thought experiments, and a large number of trains moving at implausibly high speeds, in the pages ahead of you – but our destination is physics, not maths.

It is also possible to go too far the other way, and over-stress the continuity with familiar ideas, downplaying the dislocation that the second axiom necessarily implies. I don't think this is a useful approach, because we cannot really evade the disorientation which arises from the loss of simultaneity, nor evade the necessity of thinking of time as something other than a rope along which a sequence of snapshots of 'the present' are strung.

The nature of the middle path I wish to plot out in this book is therefore, I hope, clear. I want you as the reader to appreciate SR as an account of the geometrical structure of our universe, and also see that structure as something with concrete physical consequences.

I developed the material here over a number of years, in the course of teaching an early-undergraduate course in SR at the University of Glasgow. It has therefore been read, puzzled over, commented on, and corrected by multiple cohorts of students, and the distribution of exercises in each chapter reflects both their questions and difficulties, and the inevitable requirement that I formally examine their understanding. That is, this book is not an armchair exercise in finding 'interesting topics in SR', but an attempt to bring students to the understanding of SR that they will need in their future work, and which will expand their intellectual horizons. And of course those lectures were not just a discussion group, but a course with an exam at the end of it; the presence of that exam influenced the selection and shape of material in a way which may also be of use to you.

The text here is therefore suitable to support a course: I covered the content, without appendices, in ten fairly busy lectures (and did not declare all of it examinable). But in converting the text to a book I have also had in mind both the independent reader working their way into the subject, and the student of another course who wants light shone on the landscape from a different direction. I resisted adding new core material in this process, but have expanded explanations here and there, and added various supplementary comments, either historical, mostly in discursive footnotes, or technical, in 'dangerous-bend' sections.

Although I believe my students' experience validates this text as one good route to SR, I insist that it is not the only route, nor even the only good one. Amongst the unusual things about relativity is that there exists no single royal road to an understanding of it, nor even a single obvious way of mapping the territory. Relativity always makes more sense the second time you read about it (and makes still more sense the first time you explain it to someone else), and so it is always useful to expect to read more than one introduction; to that end, I point to a number of other books in Chapter 1. My goal for this book is not to disperse, but to join, this swarm of alternatives. Throughout, I have freely referred you to other textbooks, to review articles, and to original research articles. It is not necessary to follow each of these outward-pointing links to get an understanding of the subject, but I hope they indicate the richness of the subject's connections with the rest of physics, and with the history of physics and astronomy.

It's easy for me to talk about the simplicity and minimalism of SR: you may find the claim a little surprising, if you look ahead and see a lot of solid text, including a dense undergrowth of primes and subscripts. A lot of this is *re*-explanation, however. In Chapter 5, for example, I discuss some of the 'paradoxes' of SR, deliberately pointing to results of relativistic arguments which are so unexpected as to seem surely wrong, and then going on to discuss how those results are not only consistent, but merely a different way of looking at ideas we have already encountered, so that we acquire a more textured understanding of ideas that seemed insubstantial the first time around. There is a lot of internal cross-reference knitting the book together – pointing to another version of an idea introduced previously, or presenting a preview of something done more carefully later – but I think of it as having a single short thread running from the first chapter to the last, and when you have undone and re-done these chapters in your head, along with any other broader reading, I hope you will see that thread, too.

Each chapter starts with a few high-level aims. These are collected on page xvi, and I think of them as the 'intellectual table of contents' of the book.

The exercises at the end of each chapter are, I think, important in solidifying your understanding of SR. A number of them are annotated with $d^+$, $d^-$ or $u^+$ to indicate ones which are more or less difficult, or more particularly useful, than those around them.

I have included an appendix on the link to General Relativity and gravitation. Although it is of course auxiliary to a book on Special Relativity, I believe it is useful to show that the gap between the two areas is narrower than it may at first appear, and that we can build a bridge which lands a small but significant way into the new territory. The approach I have taken to SR is designed to give this bridge as firm a foundation as possible, on the SR side.

Similarly, the appendix on relativity's contact with experiment is auxiliary but, I hope, both useful and interesting. It it not concerned with the question 'is relativity right?' – the answer to which it takes as obviously 'yes' – but instead steps back and discusses the nature of the relationship between relativity, both special and general, and experimental corroboration, and indeed has something to say about the relationship between corroboration, science in general, and scientists as a community.

Throughout the book, I have included a number of historical asides. Special Relativity does not have an unusually intricate history, but these asides are present partly because they add an extra dimension to our knowledge of the topic, but also because, by hinting at an alternative intellectual path not followed, they can colour in our understanding of the ideas as they have developed in fact. This also seems a good point at which to mention my deliberate habit of styling scientists' names, when they become adjectives, in lowercase: thus 'Newton's laws', but 'newtonian physics'. This is partly because, by the time a topic acquires an adjective like this, it has absorbed the work of a multitude of people beyond any original creator (see also 'Stigler's law'), and evading the ownership question with a lowercase letter seems both fairer and less cumbersome than a hyphenated list of all an idea's retrospectively discovered co-discoverers. Even Newton would have to go back to school if presented with a book on contemporary classical mechanics.

Throughout these notes, there are occasional sidenotes, and one or two complete sections, which are marked with a symbol like this. These 'dangerous bend' paragraphs provide supplementary detail or precision, or discuss extra subtleties, which are tangential but interesting, or which comprise extra discussion of ideas which students have in the past found confusing or easy to misunderstand. You will typically want to ignore these on a first reading, and I will sometimes presume you are re-reading, here, by referring in passing to ideas which are introduced only later in the book.

# Acknowledgements

These notes have benefitted from very thoughtful comments, criticism and error-checking, received from both colleagues and students, over the years this course has been presented, for which I am very grateful. In particular I would like to thank Richard Barrett, Andrew Conway, and Susan Stuart for detailed criticial comments when turning the notes into a book.

Thank you also to Cleon Teunissen for help with the history of the term 'Invariantz-Theorie' (p. 98, note 9); to the contributors to Wikimedia Commons for the image in Figure 1.3 (and so many resources elsewhere); and to the Historical Naval Ships Association for the scan of Figure 1.4. And thank you, finally, to the editorial staff at CUP, for their encouragement, precision, and patience.

# Aims

My goal is that you should:

1.1. understand the importance of events within Special Relativity, and the distinction between events and their coordinates in a particular frame;

1.2. appreciate why we have to define very carefully the process of measuring distances and times, and how we go about this;

2.1. understand the two axioms of SR;

2.2. appreciate the significance and inevitability of the immediate consequences of those axioms;

2.3. understand the ideas of a coordinate transformation, and of the covariance of an equation under a coordinate transformation;

3.1. understand why the concept of simultaneity is problematic in the context of SR, and how we resolve these problems;

4.1. appreciate the role of geometry in understanding spacetime, specifically the importance of the invariant interval and the Minkowski diagram;

4.2. internalise the utility of units where $c = 1$, as the natural units for discussing events in spacetime;

5.1. understand the derivation of the Lorentz transformation, and recognise its significance;

6.1. understand the concept of a 4-vector as a geometrical object, and the distinction between a vector and its components;

7.1. understand relativistic energy and momentum, the concept of energy-momentum as the magnitude of the momentum 4-vector, and conservation of the momentum 4-vector; and

7.2. understand the distinction between invariant, conserved and constant quantities.

# 1
# Introduction

*Aims*: you should:

1.1. understand the importance of events within Special Relativity, and the distinction between events and their coordinates in a particular frame; and
1.2. appreciate why we have to define very carefully the process of measuring distances and times, and how we go about this.

## 1.1 The Basic Ideas

Relativity is simple. Essentially the only *new physics* which will be introduced in this text boils down to two axioms:

1. All inertial reference frames are equivalent for the performance of all physical experiments;
2. The speed of light has the same constant value when measured in any inertial frame.

The work of understanding relativity consists of (i) appreciating what these two axioms really mean, (ii) examining their direct consequences, and (iii) thus discovering the way that we have to adjust the physics we already know. Each of these steps presents challenges.

So when I say that 'relativity is simple', I do not mean that it is easy, simply that this list of axioms is a short one. We will discover that these axioms are more subtle than may at first appear, and that they lead to conclusions which go against our usual intuitions. It is here that the difficulties arise: the maths isn't particularly hard, but we have to put a lot of effort into understanding ideas we thought were already clear, and try to think precisely about processes we thought were intuitive: what do we mean when we talk about 'the length of a stick'?

Our first step is to understand the axioms, and we'll start on that in the next chapter. What does it mean to talk about a 'transformation between frames', and why should the speed of light have such a significant place in this story?

Chapters 3 and 4 are about the reasonably direct consequences of the axioms. It's in this pair of chapters that we discover the most surprising features of SR – length contraction and time dilation – first qualitatively then quantitatively. This pair of chapters is where the most profound conceptual challenges are.

The various strands here are tied together by the main calculational tool of SR, the 'Lorentz transformation', which we derive and study in Chapter 5. This chapter is quite a long one, and detailed, so that it would be fairly easy to get lost in it. However it is really only Section 5.1 that has the new material, and the rest of the chapter is, again, an exploration of the consequences. Those consequences are often surprising, and will frequently, I think, cause you to re-read and re-think Chapters 3 or 4.

In Chapters 6 and 7 we look outwards to the rest of physics, and learn how the principles of SR oblige us to recast familiar ideas such as velocity, acceleration, frequency, momentum, mass, and energy.

Finally, in Appendix A, we look at how we can apply these ideas more generally, and discover Einstein's theory of gravity: General Relativity (GR). We can't dive too far beneath the surface here, without more advanced mathematics, but our understanding of SR will allow us to at least see the connection to the main structures of GR, and how they link to gravity.

Prior to all that, however, there are a couple of bits of terminology that it's useful to clarify in this first chapter, to avoid breaking the flow of the argument in Chapter 2. In particular:

**Events** These are things like a light-flash, or a bang, which happen at a particular place and time.
**Reference frames** These are the coordinate systems that we use to describe where and when events happen. We are concerned, in SR, with the *multiple* reference frames that may be relevant in our analysis of the physical world, and how the measurements in these frames relate to each other. We need to pay particular attention to the special case of the 'inertial frame'.
**Measuring lengths and times** We have to be quite careful about how

we measure distances, in space or time. The obvious ways of doing this contain ambiguities which can easily lead to confusion.

I expand on each of these points below. The following sections are rather short (and possibly a little dry), but we will use the ideas in them again and again and again.

It's probably a good idea to re-read these sections repeatedly as you work through the rest of the text. It possibly follows that these remarks might not make perfect sense first time; they may even at first seem absurdly over-precise, since they are making distinctions which may appear unnecessary until you have understood some of the rest of the material.

## 1.2 Events

An *event* in SR is something that happens at a particular place, at a particular instant of time. The standard examples of events are a flashbulb going off, or a banger or firecracker exploding, or two things colliding.

There is nothing *relative* about an event: if two cars crash and metal is bent, there is no 'point of view' from which the crash did not happen. Although this may seem at this point to be too obvious to be worth stating, we will discover that more things than we may expect are relative to our 'point of view', and we will use events as our way of navigating through the puzzles this produces.

We will soon discover that, although we can all agree that a particular event did happen, we might well have different answers to the questions 'where?' and 'when?' These are questions that we can answer using a *reference frame*.                                                    [Exercise 1.1]

## 1.3 Inertial Reference Frames

We need to understand first what a *reference frame* is, and then what is special about an *inertial* (reference) frame.

A *reference frame* is simply a method of assigning a position, as a set of numbers, to events. Whenever you have a coordinate system, you have a reference frame, and I will use the two terms almost interchangeably. The coordinate systems that spring first to mind are possibly the $(x, y, z)$ or $(r, \theta, \phi)$ of physics problems. Reference frames need not be fixed to a stationary body: a train driver most naturally sees the world in terms of distances in front of the train. An approaching station can quite legitimately be said to be moving – speeding up and slowing down – in the driver's reference frame.

In mechanics problems, we are used to thinking of time as a free variable: often, the point of a physics problem is to work out how the position of a thrown ball, for example, varies in time – what is $x(t)$? When thinking of events, however, it can be useful to think of them as being located using *four* coordinate numbers, $(t, x, y, z)$. We will come back to this point of view in Chapter 4.

You can generate any number of reference frames, associated with various things moving in various ways. In the context of SR, however, we can pick out some frames as special, namely those frames which are *not accelerating*.

Imagine placing a ball at rest on a table: you'd expect it to stay in place. Similarly, if you roll a ball across a table, you'd expect it to move in a straight line. This is merely the expression of Newton's first law: 'bodies move in straight lines at constant velocity, unless acted on by an external force'. In what circumstance – that is, in which frames – will this *not* be true?

Suppose you're sitting in a train which is accelerating out of a station.[1] A ball placed on a table in front of you will start to roll towards the rear of the train, rather than staying put in the way that Newton's first law seems to say it should. This observation makes perfect sense from the point of view of someone on the station platform, who sees the ball as stationary, and the train being pulled from under it. But in a reference frame attached to the train, where 'position' is perhaps measured as the distance, $x(t)$, from the rear of the train, this position will change without any force acting on the ball. We refer to the station as an 'inertial frame', and the accelerating train carriage, with respect to which Newton's law appears not to hold, as non-inertial. Similarly, if you are perched on a spinning children's round-about, and toss a ball to someone on the opposite side, it veers off to one side (interpreting this as either 'it appears to veer off to one side, from your point of view' or, more formally, 'it will be measured to veer off, as observed by someone using the rotating reference frame which is fixed to the round-about'). This motion, again, is immediately intelligible from the point of view of someone standing watching all this go on, who sees the ball go exactly where it should, but the catcher rotate out of the way. The playpark

---

[1] We're going to hear an awful lot more about this train. Although I will occasionally vary the examples by talking about rockets or boats, a train going past a station platform presents such an immediate picture of two reference frames, in constant relative motion, that it will be hard to avoid. I see no reason for wanton innovation with respect to this particular aspect of the subject.

is an inertial frame, the spinning roundabout is not. In both cases, you can tell whether you're the one in the non-inertial frame: in the first case you feel yourself pushed back into the train seat, and in the second case, it's only your grip on the roundabout that stops you flying off, pulled towards the outside by centrifugal 'force'.

Acceleration and force are intimately connected with the notion of inertial frames – an inertial frame is one which isn't accelerated in any way. From that, you would be correct to conclude that once the train has stopped accelerating, and is speeding smoothly on its way, it becomes an inertial frame again; if you closed your eyes, you wouldn't be able to tell if you were on a moving train or at rest in the station. Anything you can do whilst standing on a station platform (such as juggling, perhaps), you can also do whilst racing through that station on a train, even though, to the person watching the performance from the platform, the balls you're juggling with are moving at a hundred kilometres an hour, or so.

What we have concluded here is that, although different observers may reasonably ascribe different coordinates to events, and different speeds, there is no ambiguity about who is accelerating or not. If you are on a train picking up speed as it leaves a station, you can feel the pressure of the seat on your back, and be under no illusion that you are not moving, and there is no point of view from which the drink on the table in front of you does not look likely to spill. We will have more to say about this point in Section 2.1.

Newton's second law is more quantitative, since it relates the amount of force applied to an object, the amount it is accelerated, and the body's inertial mass, through the well-known relation $F = ma$.

We can therefore define an 'inertial frame' as follows:

*Definition of Inertial Frames:*   An *inertial reference frame* is a reference frame, with respect to which Newton's first law holds.

All this being said: don't over-think this. An inertial frame is one that isn't accelerating.

Note that, in the context of SR, inertial frames are infinite in extent; also, since all inertial frames move with constant velocity, it follows that no pair of such frames mutually accelerate.                    [Exercise 1.2]

### 1.3.1  Further Remarks on Frames ⚠

Of course, there is rather more to it than that. This definition suffices for Special Relativity, but once we consider *General Relativity* (GR) we have both the need, and the mathematical tools, for a more fundamental definition.

In brief, in GR the definition of an inertial frame is one which is in free fall, meaning one which is moving freely in a gravitational field, or freely floating, unaccelerated in interstellar space. As in SR, this is a frame in which Newton's first law still holds. This definition is locally consistent with the definition in SR, but allows us to start to discuss inertial frames which are mutually accelerating (in the specific sense that the second derivative of the separation is non-zero), such as two free-falling objects on opposite sides of the Earth. I discuss this at greater length in Appendix A.

If we acknowledge the existence of GR then, to be nit-pickingly precise, I shouldn't really talk of train carriages and station platforms as inertial frames. Firstly, there are tiny corrections due to the fact that we are on the curved surface of a rotating planet; we can ignore these in the huge majority of cases. Secondly, we should be careful when talking about throwing balls or juggling within an inertial frame, since, because of the presence of the force of gravity, a frame sitting on Earth is not inertial according to GR's stricter definition. However, as long as we are talking about SR rather than GR, as long as all the relevant motion (of inertial frames) is horizontal, and as long as no-one throws the ball further than a hundred kilometres or so (!), denying ourselves any mention of projectile motion would achieve nothing beyond removing a vivid and natural example to focus on. If you really want to, you can remove gravity from the examples by imagining the events taking place not in train carriages going through stations, but in space capsules flying past asteroids, with some suitably baroque arrangement of air jets or rockets, to supply the forces when necessary.

Also, this section is one of the few places in this text where I mention 'acceleration' (another is coming up in Section 1.6, when I talk of the 'clock hypothesis'). This is not because SR 'cannot deal with acceleration' as you might see mistakenly claimed, but simply because the novel and counter-intuitive relativistic effects, that we are going to discover in the chapters to come, are not a result of accelerating frames or observers, but instead of non-accelerating (that is, inertial) observers in mutual motion at large relative velocity. Similarly, when we (implicitly) talk of motion under gravity – such as when we talk of throwing a ball – it is perfectly reasonable for us to do so using a newtonian theory of gravity ($F = mg$) rather than anything distractingly exotic.

The mass which is the constant of proportionality in Newton's second law is what defines *inertial mass* – it indicates, roughly speaking, how resistant an object is to being accelerated by an applied force. This is distinct from the *gravitational mass* of an object, which describes how much gravitational field the object generates – in Newton's law of gravitation, $F = GMm/r^2$, both masses are gravitational masses. It turns out, however, that whatever the composition or construction of an object, its gravitational and inertial masses, though logically completely distinct, are always measured to be equal. This fact is more surprising than it may at first appear; we will examine some of its consequences in Appendix A.

## 1.4 Simultaneity: Measuring Times

How do we measure times? In SR, we repeatedly wish to talk about the times at which events happen, and more particularly the time *intervals* between events. In Chapter 3, we will discover that two observers in relative motion, who observe a pair of events, will not only ascribe different time coordinates to those events (unsusprisingly, because they are in different coordinate systems), but may also disagree about which event happens first, or whether they are simultaneous. We must therefore be careful just what we mean by 'the time of an event'.

One of the things we can hold onto in the rest of this text is that, if two events at the same spatial position happen at the same time, they are simultaneous for everybody. Einstein made this particularly clear, when he talked about what it means to assign a 'time' to an event:

> We must take into account that all our judgments in which time plays a part are always judgments of *simultaneous events*. If, for instance, I say, 'That train arrives here at 7 o'clock,' I mean something like this: 'The pointing of the small hand of my watch to 7 and the arrival of the train are simultaneous events.' (Einstein 1905)

If, however, the event happens some distance away (answering a question such as 'what time did the train pass the next signal box?'), or if we want to know what time was measured by someone in a moving frame (answering, for example, 'what was the time on the train-driver's watch as the train passed the signal box?'), things are not so simple, as most of the rest of this text makes clear. Special Relativity is very clear about what we mean by 'the time of an event': when we talk about the time of an event, we *always* mean the time of the event *as measured on a clock carried by a local observer*, that is, an observer at the same spatial position as the event (which is rather unfortunate if the event in question is an explosion of some type – but what are friends for?), who is stationary with respect to the frame they represent. We will typically imagine more than one observer at an event; indeed we imagine one local observer per frame of interest, stationary in that frame, and responsible for reporting the space and time coordinates of the event as

**Figure 1.1** Our observers, equipped with their clocks and surveyors' wheels.

measured in that observer's frame.[2]

What we never do is have an observer in one location measure and report the time of an event at another location. To do so, we'd have to concern ourselves with a number of complications, most prominently correcting for the time of flight of the light from the event to our observer's eye, which depends on the distance between them, and so on. It would be possible to be clever and correct for these, but we simply avoid doing so by exclusively using local observers.

We suppose that a frame has a plentiful supply of observers, and that we can position them, and their clocks, wherever we need to make an observation (we're going to take 'observation' and 'measurement' to be synonyms here). Or you can imagine observers positioned *everywhere* in a frame, ready to take note of any events which happen next to them. Our observers – see Figure 1.1 – are both lazy and very short-sighted: once they have established their location they never move, and they *will only ever observe events which happen right in front of them*. When an event happens, they note the time on their clock, ready to report it along with their measured position when required.

Since the observers never move, the frame they are standing in is special to them, it is the *rest frame* of the observer. At the risk of belabouring the point, if we have two observers, one on a train and one on a station platform, then the rest frame for one observer is the frame attached to the station (within which neither the station nor the observer are moving), and the rest frame for the other is attached to the train (within which the train is not moving, but the station is).

The observers in a particular frame have one further property of impor-

---

[2] Note that it makes no sense to talk of being 'stationary with respect to an event' or to talk of 'the rest frame of an event': since an event is instantaneous, it cannot be said to be moving in any frame. This also means that there is no observer who is 'special' with respect to an event.

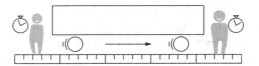

**Figure 1.2** Measuring the length of a train, using a ruler painted on the edge of the station platform. The observers are standing on the platform. You should imagine observers all along the platform; I've shown only the two who happen to be next to the ends of the train at the observation time.

tance: as well as being stationary in the frame, we presume that they have arranged things so that the clocks they carry are, and remain, mutually synchronised. I have a little more to say about this in Section 2.2.1, but I mention it here only to reassure you that this is not on the list of unexpectedly complicated things.

Although an event might have a number of adjacent observers, in different frames, this doesn't imply that all observers report the *same* time. The observers' watches may differ for trivial reasons – perhaps their watches are set to different time zones, or designed to run at different rates. Or, less trivially, they may be running at different rates for relativistic reasons that we'll come to later. We assume that, despite these complications, all of the clocks tick out a time – produce a *number* – which is linearly related to the passage of time. [Exercise 1.3]

## 1.5 Simultaneity: Measuring Lengths

How should we measure the length of a moving object such as the train carriage of Figure 1.2? The obvious method is some variant of laying out a metre stick or measuring tape, and looking at how the markings line up with the moving carriage. But doing so raises some awkward questions. I said above that we want to avoid making any non-local measurements, and make all observations with co-located observers. How do we go about that in this situation? You obviously have to measure both ends of the object simultaneously: does that 'simultaneously' depend on where you're standing relative to the object? There are questions here which are only apparently straightforward, and to which SR provides surprising answers.

The way we measure lengths and times in SR is therefore as follows. We position observers at strategic points in the reference frames of interest. We can know these observers' coordinates in one frame or another (I find it useful to imagine the $x$-axis painted on the platform edge). The observers make records of events which happen at their location, and afterwards compare notes and draw conclusions. Specifically, you would measure the 'length of a train' by subtracting the coordinates of the two observers who observed opposite ends of the train at a pre-arranged time (remember that we are assuming that all of the observers in a particular frame have pre-synchronised clocks). We return to this in Chapter 3.

If these two observers are inside the train carriage, stationary at opposite ends, then they get the value you would intuit: this is merely a complicated way of measuring the length they could obtain by stretching a tape-measure between them. For the case of the observers on the platform measuring the moving train, this does initially seem a complicated way of organising things, but it has the virtue of being unambiguous about when things happen and where.

To summarise, this approach relies on three things.

1. It requires a specific procedure for synchronising clocks, which is described briefly in Section 2.2.1.
2. It assumes that there is no ambiguity about two events at the same position and time being regarded as simultaneous. This has to be true: the fact that two cars crash – because they were in the same position at the same time, and so are attempting to occupy the same space simultaneously – cannot depend in any way on your point of view.
3. We assume that moving clocks measure the passage of time accurately – that they are 'good clocks' in the specific sense of the 'clock hypothesis' (see Section 1.6).

The second point in this list does need saying, possibly surprisingly: we will later discover that the simultaneity of two events not at the same position *does* depend on your point of view.

## 1.6 The Clock Hypothesis

The *clock hypothesis* is the assumption that there is nothing intrinsic to motion, or to acceleration, which stops a clock being a reliable measure of the passage of time.

The first (rather trivial) part of this is the presumption that, whatever the technology we're using for our clocks, it's appropriate for the situation, so that we're not, for example, using a pendulum clock in space or on board ship, or a clock which will snap if we accelerate it. A very concrete example of a 'good clock' is an atomic clock, which depends on fundamental physics

**Figure 1.3** A Han-dynasty horse-drawn odometer. With bells on. (Image: stone rubbing of an AD 125 tomb carving; Wikimedia commons.)

to mark out a timescale.

Secondly, and much more significantly, the hypothesis asserts that there is nothing about acceleration which necessarily affects the way a clock measures the passage of time; an accelerating clock will 'tick' at the same rate as an unaccelerated clock moving at the same instantaneous speed.[3]

Put another way, this is the hypothesis that there are no additional complications we need worry about here, and that as far as we are concerned, time *is* the thing showing on the face of a clock. We will occasionally talk of a *good clock* – we are invoking the clock hypothesis when we do so.

An odometer, or a surveyor's measuring wheel, is a device which records how far a vehicle, or a surveyor, has travelled by recording how much the device's wheel has turned. This is a very direct measurement of the distance travelled through space (though not quite as direct as the idea of laying out a measuring rod end-to-end). These need be no more complicated than a wheel on a stick, but it's possible to make them more elaborate (see Figure 1.3 for a spectacular example).

A useful corresponding image, I find, is to imagine a clock to be like an old-fashioned 'taffrail log' (Figure 1.4), which is type of propellor towed behind a boat, which records how much water the boat has moved through. It's like an odometer for water. Analogously, a clock is a device which records how much time the clock has been dragged through.

A further way of putting this, which will mean more to you after you have looked at Chapter 5, is to say that proper time is the measure of the length of a worldline, and that this length is a property of the worldline, as

---

[3] In a famous experiment, Bailey et al. (1977) measured the decay rates (which mark the passage of time, and thus count as a clock for this purpose) of muons in a storage ring at CERN; because the particles were moving in a circle, they were subject to very high centripetal accelerations. The muons decayed at the same rate as they would if they were moving at the same speed in a straight line.

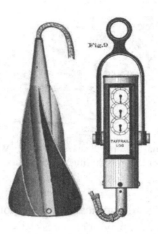

**Figure 1.4** A 'taffrail log': the propellor is towed through the water, and the number of turns is recorded on the dials (from Plate 6 of S. B. Luce, *Textbook of Seamanship* (1891); image courtesy Historical Naval Ships Association).

a path through spacetime, rather than being a property of any physical system *within* spacetime, such as a clock. In these terms, the clock hypothesis is the idea that a 'good clock' is a clock which, though it is a physical thing within spacetime, faithfully records this worldline length.

## 1.7 Standard Configuration

Finally, a bit of conventional terminology to do with reference frames. Two (inertial reference) frames $S$ and $S'$, with spatial coordinates $(x, y, z)$ and $(x', y', z')$, respectively, and time coordinates $t$ and $t'$ are said to be in *standard configuration* if:

1. they are aligned so that the $(x, y, z)$ and $(x', y', z')$ axes are parallel;
2. the frame $S'$ is moving along the $x$-axis with velocity $v$; and
3. we set the zero of the time coordinates so that the origins coincide at $t = t' = 0$; this means that the origin of the $S'$ frame (which of course remains at position $x' = 0$ by definition) is always at position $x_{S'} = vt$ in frame $S$.

This arrangement, shown in Figure 1.5, makes a lot of example problems somewhat easier, and is the convention assumed by the 'Lorentz transformation' which we will meet in Chapter 5.

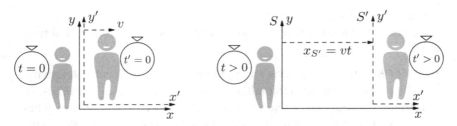

**Figure 1.5** Standard configuration: two frames $S$ and $S'$, where frame $S'$ is moving at speed $v$ with respect to frame $S$; the axes coincide at time $t = t' = 0$.

In Figure 1.5, *both* of the observers are at positions which can be given coordinates in both frames, $(t, x, y, z)$ and $(t', x', y', z')$. The left-hand observer in each figure is at rest in frame $S$, meaning that they stay stationary at position $(x, y, z)$, but of course move through time $t$, as shown on their clock; their position in terms of coordinates $(x', y', z')$ *changes*, with $dx'/dt' = -v$. The other observer, at rest in frame $S'$, stays at coordinates $(x', y', z')$, but with changing $x$-coordinate: $dx/dt = v$.

When we refer to 'frame $S$' and 'frame $S'$', we will interchangeably be referring either to the frames themselves, or to the sets of coordinates $(t, x, y, z)$ or $(t', x', y', z')$. It's also worth mentioning in passing that, in standard configuration, we can presume that the corresponding coordinates in the two frames, such as $t$ and $t'$, or $x$ and $x'$, will always be in the same units, whether these be nanoseconds or parsecs.

In standard configuration, the motion of the frames is always along the $x$- and $x'$-axes, and in the examples we use we can always choose the axes so that this is so. This is not a fundamental restriction, and we can relax it at the cost of mere algebra. There is a little more to say about this in Section 5.2.1.

## 1.8 Further Reading

When learning relativity, even more than with other subjects, you benefit from hearing or reading things multiple times, from different authors, and from different points of view. I mention a couple of good introductions below, but there is really no substitute for going to the 'physics' section in the library, looking through the SR books there, and finding one which makes sense to *you*.[4]

I recommend in particular the books listed below, which excel in different ways.

- Taylor & Wheeler (1992) is an excellent account of Special Relativity, written in a style which is simultaneously conversational and rigorous. It introduces the subject from a geometrical perspective from the very beginning, which makes it quite natural to make links to General Relativity.
- Rindler (2006) contains a clear account of the subject, using a broadly similar tack to my own. Rindler takes great pains to confront and explain the subtleties involved in the subject, and explains things both carefully and lucidly; he is also much less geometrical in tone than Taylor & Wheeler. I think that both Rindler and Taylor & Wheeler see SR essentially as a prologomenon to GR, and are consequently admirably careful in their treatment of SR, to the extent that in Rindler's book the distinction between them is rather porous. Because of this, I don't think Rindler is a perfect first introduction to SR, but he is invaluable as a second source, to turn to after a first source leaves you uncertain of a subtlety, or simply wanting more detail. The second half of this book is about GR, and goes substantially beyond the scope of this text (and he takes what is now rather an old-fashioned approach).[5]
- French (1968) is a justly well-known account, which has been in print since it was first published. It takes a rather traditional approach to the subject. Here, the motivation for SR is particle physics, so it talks a lot about dynamics, and is not much interested in geometry. In terms of style and approach, I think of it as the polar opposite of Gourgoulhon (2013), and enjoy both.
- Barton (1999) is another book with in some ways a rather traditional approach, though more modern than French. Its particular strength is that, like Rindler, it explains things very carefully and with scrupulous reference to background material. This makes it slightly hard work as an introduction to SR, but very valuable to me when it allows me to direct you towards supporting details, when they might break the flow below.
- Takeuchi (2012) is a charming book which introduces the key ideas of SR purely through diagrams, without any equations. It overlaps comfort-

---

[4] There is an annotated and updated list of recommendable texts on relativity at https://math.ucr.edu/home/baez/physics/Administrivia/rel_booklist.html.

[5] Note that Rindler (1977) and Rindler (2006) are different books, even though they are by the same author, and have almost identical titles. They are so similar that the latter almost seems like a (significantly) revised edition of the former. I will refer to the 2006 book below, but if you happen to have access only to the 1977 book, it is as rich in insight.

ably with the material up to Chapter 4, and touches on the paradoxes in Chapter 5.

*Notation*: If you refer to a range of books (and, to be clear, I think you should), then you should be aware that there are a few notational variants between texts. In particular, see the notes regarding this in Section 4.4 and Section 6.2.

More advanced textbooks, on General Relativity, often include a very compact account of Special Relativity in their initial chapters. It can be useful to look at these once you have a basic understanding of the topic, as this can often highlight the key features, and show you how 'simple' SR is, in the slightly specialised sense I mentioned at the beginning of this chapter.

- Schutz (2009) is a textbook on *General* Relativity, and is therefore generally beyond the scope of this text. However the first chapter gives a breakneck account of SR, and the second a similarly abstract account of vector analysis, which were influential on Chapter 6.
- Landau & Lifshitz (1975) covers a daunting range of classical physics, but Chapter 1 covers SR in a particularly compact way, which might help solidify your understanding *after* you have broadly understood the material.
- Taylor et al. (2019) is a book on GR by the same authors as Taylor & Wheeler (1992). It has an unusual style, but is just as vivid. As of 2021 the book is still only in draft, and available online.
- I should also mention the book Gray (2019).

There are also some books which might provide some historical insight and context. It's worthwhile taking at least a look at Einstein's famous 1905 paper introducing the ideas which were later termed 'Special Relativity' (Einstein 1905). The first few sections are worth reading for their clarity and directness (and I'm going to quote from this paper in the next chapter). That said, this paper is of historical interest rather than being a great introduction because, amongst other things, the notation is somewhat different from what we would use today. It is also unusual, amongst foundational papers in physics, for being at least intelligible; most analogous papers are for historical specialists. Einstein's own popular account of relativity (Einstein 1920) is very readable, though it's naturally a little old-fashioned in places. You can see the influence of this book, and its examples, in many later SR textbooks. *The Principle of Relativity* (Lorentz et al. 1952) is a collection of (translations of) original papers on the Special and General theories, including Einstein's 1905 paper, but also some earlier papers by Lorentz suggesting interpretations of the Michelson–Morley experiment.

Other books which have interesting notes on SR, but which should be regarded as strictly supplemental, include: David Mermin's book (1990) is a (very good) collection of essays, covering a wide variety of topics from the practice of physics, through quantum mechanics, to relativity; Chapters 19 to 21 are about unusual approaches to the teaching of relativity. Another book to mention, just because I like it, is Malcolm Longair's collection of lecture notes and case studies, covering several areas in theoretical physics (Longair 2020): it pulls no mathematical punches, but is full of insights, including chapters on SR and GR.

There are several popular science books which are about, or which mention, relativity – these aren't to be despised just because you're now doing the subject 'properly'. These books tend to ignore any maths, and skip more careful detail (so they won't get you through an exam), but in exchange they spend their efforts on the underlying ideas. Those underlying ideas, and the development of your intuition about relativity, are things that can sometimes be forgotten in more formal texts. I've always liked Schwartz & McGuinness (2012), which is a cartoon book but very clear, and which has a special place in my affections as the book which first introduced schoolboy me to the ideas of Special Relativity.

## Exercises

### Exercise 1.1 (§1.2)

Which of the following are *events*?

1. A supernova explosion.
2. A concert.
3. The whole country clapping hands at once.
4. A collision between two particles in the LHC.                    $[d^-]$

### Exercise 1.2 (§1.3)

Which of the following represent inertial frames?

1. A plane, while cruising at altitude.
2. A plane, while landing.
3. The surface of the Earth.
4. A stationary car.

5. A car moving at a straight line at a constant speed.
6. A car cornering at a constant speed.
7. A stationary lift.
8. A free-falling lift (this last one is rather subtle and relates to the dangerous-bend paragraphs at the end of Section 1.3; don't spend too much time thinking about this, at least until you've looked at Appendix A).    [ $d^-$ ]

## Exercise 1.3 (§1.4)

Imagine you're standing on a bridge over a motorway, looking at a traffic cop standing beside the road. A car driver sneezes precisely at the point when they're passing the police officer. Whose watch do you consult to find out the time of this sneeze in your frame (presuming all this is happening at relativistic speed)?

1. Your own watch.
2. The police officer's watch.
3. The driver's watch.    [ $d^-$ ]

# 2

# The Axioms

In this chapter, I am going to introduce the two axioms of Special Relativity. These axioms are, to an extent, the only new *physics* introduced in this text: once I have introduced them and made them plausible, the rest of our work is devoted to examining their consequences, and the way in which they change the physics we are already familiar with.

*Aims*: you should:

2.1. understand the two axioms of SR;
2.2. appreciate the significance and inevitability of the immediate consequences of those axioms; and
2.3. understand the ideas of a coordinate transformation, and of the covariance of an equation under a coordinate transformation.

## 2.1 The First Postulate: the Principle of Relativity

A form of the *Principle of Relativity* (or *Relativity Principle*, RP) was described very clearly by Galileo, and his account is both clear and charming enough to quote at length:

> SALVATIUS: Shut yourself up with some friend in the main cabin below decks on some large ship, and have with you there some flies, butterflies, and other small flying animals. Have a large bowl of water with some fish in it; hang up a bottle that empties drop by drop into a wide vessel beneath it. With the ship standing still, observe carefully how the little animals fly with equal speed to all sides of the cabin. The fish swim indifferently in all directions; the drops fall into the vessel beneath; and, in throwing something to your friend, you need throw it no more strongly in one direction than another, the distances being equal; jumping with your

feet together, you pass equal spaces in every direction. When you have observed all these things carefully (though there is no doubt that when the ship is standing still everything must happen in this way), have the ship proceed with any speed you like, so long as the motion is uniform and not fluctuating this way and that. You will discover not the least change in all the effects named, nor could you tell from any of them whether the ship was moving or standing still. In jumping, you will pass on the floor the same spaces as before, nor will you make larger jumps toward the stern than toward the prow even though the ship is moving quite rapidly, despite the fact that during the time that you are in the air the floor under you will be going in a direction opposite to your jump. In throwing something to your companion, you will need no more force to get it to him whether he is in the direction of the bow or the stern, with yourself situated opposite. The droplets will fall as before into the vessel beneath without dropping toward the stern, although while the drops are in the air the ship runs many spans. The fish in their water will swim toward the front of their bowl with no more effort than toward the back, and will go with equal ease to bait placed anywhere around the edges of the bowl. Finally the butterflies and flies will continue their flights indifferently toward every side, nor will it ever happen that they are concentrated toward the stern, as if tired out from keeping up with the course of the ship, from which they will have been separated during long intervals by keeping themselves in the air.

Galileo Galilei (1632), quoted in Taylor & Wheeler (1992, §3.1)

This is a very vivid account of the Relativity Principle (RP), which I shall state more precisely at the end of this section. It's also an illustration of the idea of an inertial frame, which we discussed in Section 1.3. Another way of phrasing the principle is that 'you can't tell if you're moving' – there's no experiment you can do which would allow you to distinguish between a moving and a stationary frame.

The Relativity Principle as quoted here, and discussed below, is also sometimes referred to as the Equivalence Principle (for SR). However, when you go on to study GR, you will discover that it has an Equivalence Principle of its own (and if one is reading a careful account, one discovers even weak, strong, and semistrong variants of it). To avoid confusion, it seems best to leave the term '*The* Equivalence Principle' to GR: the RP is deeply linked to the Equivalence Principle, so the issue here is more one of terminology than physics. The various equivalence principles are discussed at illuminating length in Rindler (2006, ch. 1).

Galileo's Relativity Principle implicitly refers only to mechanics. However, given that you need mechanical components to do any electromagnetic experiment, and given that all mechanical objects are held together by (atomic) electromagnetic forces, it would seem unavoidable that it must

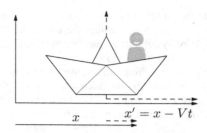

**Figure 2.1** An observer in a boat, at position $x'$ in the moving frame, and position $x$ in the harbour frame.

apply to electromagnetism as well. And since we are made up from the same atoms as everything else, it seems inevitable that the principle also extends to biology. See also Einstein's remarks quoted later in this section, and Barton (1999, ch. 3); I have more to say about this in Section 5.8.2.

From the RP, one can show that, with certain obvious (but, as we will discover, wrong) assumptions about the nature of space and time, one could derive the *galilean transformation* (GT), which relates two frames in standard configuration (see Section 1.7 or Figure 2.1):

$$
\begin{aligned}
x' &= x - Vt \\
y' &= y \\
z' &= z \\
t' &= t.
\end{aligned}
\tag{2.1a}
$$

You have possibly seen this expression before, though probably not in this form. This transformation relates the coordinates of an event $(t, x, y, z)$, measured in frame $S$, to the coordinates of the *same* event $(t', x', y', z')$ in frame $S'$. Differentiating these, we find that

$$
\begin{aligned}
v'_x &= v_x - V \\
v'_y &= v_y \\
v'_z &= v_z,
\end{aligned}
\tag{2.1b}
$$

and differentiating again

$$
a' = a,
\tag{2.1c}
$$

where $v_x = dx/dt$ is the $x$-component of velocity, and so on.

If we take the RP seriously (and we should, of course), it tells us that any putative law of mechanics which *does* appear to allow you to distinguish between reference frames cannot in fact be a law of physics. This means that numbers might change from frame to frame – if you walk along a moving train, you are moving faster with respect to the nearby platform than you are moving with respect to the other passengers – but the physics doesn't change. If you can juggle on the platform, you can do so on the (smoothly) moving train as well. If you throw a ball whilst on a moving train, the usual constant-acceleration equations tell us that it will follow a parabola, with certain parameters of maximum height, angle, and so on; someone watching this ball from a station platform *also* sees a parabola: the ball is moving at a different speed and a different angle, and it moves a greater distance – it is a different parabola – but it remains a parabola nonetheless. The physics has led us to solutions for the path of the ball which are different in the two cases, but only to the extent of them being merely two variants of the same shape.

How do we phrase this last paragraph in equation form? Consider Newton's second law, in the alternative form $F = dp/dt$ – that is, we conceive of force as a thing which changes the momentum of an object it acts on, which only secondarily results in acceleration. Specifically, consider motion under gravity, so that $F = mg$. Writing $p = mv = mdx/dt$, and $p' = mv'$ in the primed frame,

$$
\begin{aligned}
p' = mv' &= m\frac{dx'}{dt'} \\
&= m\frac{d}{dt}(x - Vt) \qquad \text{using Eq. (2.1a)} \\
&= p - mV
\end{aligned}
$$

(that is, we are differentiating the $x'$-coordinate with respect to the time-coordinate in that frame, and not merely sticking primes on the expression). Thus

$$
mg = dp/dt \Leftrightarrow mg = dp'/dt'.
$$

It doesn't matter what coordinates we use – i.e., it doesn't matter which frame we're in – the laws of physics are the same.

Fundamental physical laws take the same form in different frames, which means that their predicted consequences take the same form as well. Consider the constant-acceleration equation

$$
x = v_0 t + \frac{1}{2}at^2. \tag{2.2}
$$

Transform this to the moving frame by replacing unprimed quantities by primed ones using Eq. (2.1), and we find

$$x' = v_0' t' + \frac{1}{2} a' t'^2. \qquad (2.3)$$

That is, we find exactly the same relation, as if we had simply put primes on each of the quantities in Eq. (2.2). This is known as 'form invariance', or sometimes *covariance*, and indicates that the expression Eq. (2.3) has exactly the same *form* as Eq. (2.2), with the only difference being that we have different numerical values for the coefficients and coordinates (in general, though, $a' = a$ and $t' = t$ according to the GT). Barton (1999, §2.3.3) discusses this usefully; see also Exercise 2.4.

Going further, if all frames are equivalent, in the sense of the RP, then there is no frame that is special, and in particular this means that we cannot identify any frame corresponding to a state of absolute rest. But that in turn means that the very idea of such a frame is redundant.

We can generalise this, and say that the RP, *in classical mechanics*, demands that all laws of mechanics, and by obvious extension all other physical laws, be *covariant under the galilean transformation*. Though you probably wouldn't naturally phrase things like this, this is entirely in accordance with your (and my) physical intuition, and it seems amply corroborated by the majority of our experience. Writing down Eq. (2.1) seems little more than an exercise in notation.

Note carefully that, although we are talking in this section about 'the Principle of Relativity', we are *not* yet talking about 'Special Relativity'. The RP is consistent with both newtonian physics and SR: it is the second axiom, in Section 2.2 below, which distinguishes between them (I have more to say about this in Section 5.8.1, which is a 'dangerous bend' section).

This first axiom is consistent with our intuition, with our mathematical tastes, and also, it seems, with experiment.

Everything, therefore, seems to be rosy.                    [Exercises 2.1–2.3]

### 2.1.1 Electromagnetism

Everything, in fact, *was* rosy, until the end of the nineteenth century. Around then, physicists were investigating *Maxwell's equations*, one of the highpoints of nineteenth-century physics, which unified all of the phenomena of electricity and magnetism into a single formalism of tremendously insightful power and overwhelmingly successful application. For concreteness, let's remind ourselves what Maxwell's equations look like:[1]

$$\nabla \cdot \mathbf{E} = \frac{\rho}{\epsilon_0}$$

$$\nabla \cdot \mathbf{B} = 0$$

$$\nabla \times \mathbf{E} = -\frac{\partial \mathbf{B}}{\partial t} \tag{2.4}$$

$$\nabla \times \mathbf{B} = \mu_0 \left( \mathbf{J} + \epsilon_0 \frac{\partial \mathbf{E}}{\partial t} \right).$$

These four equations are essentially Gauss's laws for the electric and magnetic fields $\mathbf{E}$ and $\mathbf{B}$, Faraday's law, and Ampère's law, in terms of the charge density $\rho$ and current $\mathbf{J}$. Maxwell's contribution (amongst many others) was to realise that these four different things, extensively studied in the nineteenth century, were facets of a single underlying structure. As such, Maxwell's equations were as amply corroborated as the various precursor laws, and were used to show that electromagnetic phenomena from radio waves to visible light were merely different views of a single thing, and that they propagate at a single speed, the speed of light.

Maxwell's equations work.

Unfortunately, they are not invariant under a galilean transformation. The wave equation, and Maxwell's equations, do not transform into themselves under a GT.

> This is fairly easy to show for the wave equation, slightly more involved for Maxwell's equations. We have more to say about this in Section 5.8.2. More advanced textbooks on electromagnetic theory also tend to have sections on SR, which make this point more or less emphatically.

It appeared that Maxwell's equations had their simplest form – that is, Eq. (2.4) – only in a frame which was *not moving*. The fact that the equations of electromagnetism are not invariant under the GT appeared to indicate that, whenever you watched an electromagnetic experiment (such as an ammeter, or a microwave oven) in a moving frame, it should work differently from that same experiment in a stationary frame. Specifically, it suggests that there actually exists such a unique absolutely stationary frame, which is otherwise rendered unnecessary by the RP.                    [Exercise 2.4]

---

[1] If you haven't encountered Maxwell's equations yet, or if this notation is unfamiliar to you, don't worry – the point here is to show that, excepting some notational complexities (OK, quite a few complexities), they are as elegant and simple as Newton's laws. The $\nabla$ symbol on the left hand side is a differential operator, so that this is a set of differential equations relating spatial gradients in the electric and magnetic fields to distributions of charge and current, and temporal changes in the magnetic and electric fields.

## 2.1.2 Einstein, Electromagnetism, and the Foundations of Special Relativity

Einstein noted that electrodynamics appeared to be concerned only with relative motion, and did not take a different form when viewed in a moving frame. His 1905 paper is very clear on this point, and begins:

> It is known that Maxwell's electrodynamics – as usually understood at the present time – when applied to moving bodies, leads to asymmetries which do not appear to be inherent in the phenomena. Take, for example, the reciprocal electrodynamic action of a magnet and a conductor. The observable phenomenon here depends only on the relative motion of the conductor and the magnet, whereas the customary view draws a sharp distinction between the two cases in which either the one or the other of these bodies is in motion. For if the magnet is in motion and the conductor at rest, there arises in the neighbourhood of the magnet an electric field with a certain definite energy, producing a current at the places where parts of the conductor are situated. But if the magnet is stationary and the conductor in motion, no electric field arises in the neighbourhood of the magnet. In the conductor, however, we find an electromotive force, to which in itself there is no corresponding energy, but which gives rise – assuming equality of relative motion in the two cases discussed – to electric currents of the same path and intensity as those produced by the electric forces in the former case. (Einstein 1905)

The puzzle that Einstein is drawing attention to is that, although there are two significantly different explanations of what is happening, when either the magnet or the conductor is in motion, the *observable* current is identical. The thing that is special about Einstein's approach here is that he sees this, not as a curiosity, but as a massive problem: *why* is there this unexplained symmetry? What is it telling us?

Another, linked, problem was that of the aether. Since light is an electromagnetic wave, it seems obvious that, like water waves or sound waves, there must be something that light waves propagate *in*. This 'light medium' was named the aether, and had the apparently contradictory properties of being both very rigid (so that it could sustain the very high frequencies of light) and very tenuous (so that objects such as planets could move through it freely). The aether is an obvious candidate for the frame of absolute rest.

The Earth moves around the Sun in its orbit, with a constantly changing velocity. It followed, therefore, that there was some point in its orbit at which it had a maximum, and another point at which it had its minimum, speed with respect to the putative aether. Although this speed is rather slow compared to the speed of light, it should have been possible to *measure* the change in the velocity of the Earth with respect to the aether or the absolute rest frame. There was therefore a series of experiments in the late nineteenth and early twentieth centuries which attempted to measure this phenomenon: the Michelson–Morley aether-wind experiment attempted to measure the different light-travel times for beams directed along and across the flow of the aether; the Fizeau experiment and Lodge's experiments attempted to detect the extent to which the aether could be dragged along by fast-moving objects on Earth. All of them failed: no-one was able to detect the Earth's movement through the aether, or the movement relative to the absolute rest frame, which the galilean transformation and the apparently necessary properties of an apparently necessary aether demanded.

The aether-wind experiments are discussed in most relativity textbooks. French (1968), for example, gives clear accounts. I have quite a lot more to say about this in Appendix B.

Thus Maxwell's equations appear to be inconsistent with the galilean transformation (in the sense that Eq. (2.4) is not form-invariant under the GT), or else the equations are inconsistent with the relativity principle (in the sense that they appear to point towards the existence of a special reference frame). At this stage there were a number of options.

(i) Perhaps Maxwell's equations were wrong. Although the unification that Maxwell's equations represent is intellectually satisfying, and although they appear to precisely match experiment, perhaps there was some further careful experiment which could be devised which would uncover some contradiction between theory and experiment.

(ii) Perhaps the Relativity Principle was wrong, although one possible result of the repeated failures to measure the Earth's motion through the aether could, I imagine, have been the weakening of the idea of an absolute rest-frame.

(iii) Perhaps the GT was wrong, though this transformation seems so obvious, and so bound up with our other preconceptions that it would be difficult to see how this would be possible.

(iv) Perhaps there was some further physics at work. There were suggestions that an experimental apparatus, or even the Earth itself, might be able to drag the aether along with it, enough to wipe out any detection from the Michelson–Morley experiment. Lorentz managed to find an alternative to the GT – now known as the Lorentz or FitzGerald–Lorentz transformation[2] – under which Maxwell's equations *are* covariant, but he and the rest of the

---

[2] For more about Lorentz, FitzGerald, and this transformation, see Section 5.8.2.

community were then left with the problem of explaining why electromagnetism was apparently uniquely subject to a different transformation law from everything else. The Lorentz transformation appeared to indicate that objects would change their lengths, and time be distorted, when moving head-on into the aether, without there being any clear physical mechanism for this. It would have been clear that something was very wrong.[3]

This is a fascinating episode in the history of science, but the resolution was that (i) Maxwell's equations are right, (ii) the Relativity Principle is right, (iii) the galilean transformation is only approximate, and (iv) Special Relativity is the new physics to come out of this.

By saying that the GT is 'approximate' I mean that, although the transformation clearly works in our normal experience, we can find circumstances (namely when we are moving quickly, at 'relativistic speed') where it produces wrong predictions.

Einstein explains this as clearly as anyone. In his 1905 paper which introduced SR ('On the Electrodynamics of Moving Bodies'), he opens with the paragraph above commenting that only *relative* motion is important in Maxwell's equations, and then goes on to say, with magisterial finality:

> Examples of this sort, together with the unsuccessful attempts to discover any motion of the Earth relatively to the 'light medium', suggest that the phenomena of electrodynamics as well as of mechanics possess no properties corresponding to the idea of absolute rest. They suggest rather that, as has already been shown to the first order of small quantities, the same laws of electrodynamics and optics will be valid for all frames of reference for which the equations of mechanics hold good. We will raise this conjecture (the purport of which will hereafter be called the 'Principle of Relativity') to the status of a postulate... (Einstein 1905)

In other words, he is saying that Galileo's Relativity Principle, which had really only been a statement about mechanical experiments, now applied to

---

[3] There are some interesting historical details in Einstein's 'Autobiographical Notes' (1991) and Pais's scientific biography of Einstein (2005, ch. 6). The apparent non-covariance of Maxwell's equations, and the problems with the aether theory, were well-recognised problems by the end of the nineteenth century, and the historically interesting thing here is the extent to which Einstein was aware of the prior work but didn't feel he needed to build on it, since his motivations for the 1905 paper were largely philosophical. Lorentz's discussion of the Michelson–Morley experiment describes how one might reconcile it with the aether theory, by assuming inter-molecular forces are modified by the aether in a particular way 'though to be sure, there is no reason for doing so' (Lorentz 1895, §4); and Bell (2004, ch. 9) has described a potential way of teaching relativity, by considering the electrostatic field of a point charge, which would now be regarded as extremely eccentric, but which would have made a lot of sense to Einstein's contemporaries, and which is illuminating therefore (and see Section 5.8.2 below). It was clear in 1905 that the physics of mechanics and of electromagnetism were intimately related to one another, so that the fact they seemed to observe incompatible transformation laws was a major anomaly.

*all* of physics.

We can recast the Principle of Relativity (RP), the first postulate of Special Relativity, as follows:

> *The Principle of Relativity:*   All inertial frames are equivalent for the performance of *all* physical experiments.

There is no physical (or chemical or biological or sociological or musical) experiment I can do which will have a different result when I'm moving uniformly from when I'm stationary. There is therefore no need for even the idea of a standard of absolute rest. And further (but importantly for Einstein's approach) any theory which appears to require such a standard, must be wrong in principle.

Put another way, although Newton devised his laws of mechanics in an 'absolute rest frame', and although Maxwell devised electromagnetism in an 'aether frame', the RP says that the same physical explanation will be good in *any* inertial frame. It may not be obvious, but the demand that this be so puts a severe constraint on the theories in question, or on the relationships we must find between inertial frames. [Exercise 2.5]

## 2.2 The Second Postulate: the Constancy of the Speed of Light

The passage I quoted above, from Einstein's 1905 paper, goes on to say:

> We will raise this conjecture (the purport of which will hereafter be called the 'Principle of Relativity') to the status of a postulate, and also introduce another postulate, which is only apparently irreconcilable with the former, namely, that light is always propagated in empty space with a definite velocity *c* which is independent of the state of motion of the emitting body. (Einstein 1905)

This is the second postulate of Special Relativity.

This doesn't seem particularly remarkable at first reading; after all, we *know* that 'the speed of light' is one of nature's fundamental constants, at $c = 299\,792\,458 \text{ m s}^{-1}$. The sting is in the final remark, 'independent of the state of motion of the emitting body'. At first thought, there would seem to be three things that 'the speed of light' could mean:

1. the speed relative to the emitter (like a projectile);
2. the speed relative to the transmitting medium (like water or sound); or
3. the speed relative to the detector.

Option 2 is ruled out by the first postulate: if this were true then the frame in which light had this special value would be picked out as special; the RP also incidentally excludes the notion of the aether.

Option 1 also turns out not to be the case, if the statement here is taken to mean that light behaves just like a classical projectile. While light is always emitted at the speed $c$, option 1 suggests that it may potentially arrive at a different speed at a moving detector. This is what we would intuit from the velocity addition part of the GT, Eq. (2.1b), which says that velocities add in a straightforward way; this turns out not to be true for light, or indeed any object moving at a significant fraction of the speed of light.

No, option 3 is the case, so that, no matter what sort of experiment you are doing, whether you are directly observing the travel-time of a flash of light, or doing some interferometric experiment, the speed of light relative to your apparatus will always have the same numerical value. This is perfectly independent of how fast you are moving relative to the source: it is independent of whichever inertial frame you are in, so that another observer, measuring the *same* flash of light from their moving laboratory, will measure the speed of light relative to *their* detectors to have exactly the same value.

*Constancy of c:*   There exists a finite constant speed

$$c = 299\,792\,458 \text{ m s}^{-1},$$

such that anything which moves at this speed in one inertial frame is measured to move at that speed in all other inertial frames.

Barton (1999, §3.1) gives a wonderfully careful expression of this. There is no real way of *justifying* this postulate: it is simply a truth of our universe, and we can do nothing more than simply demonstrate its truth through experiment.

This experimental corroboration might take the form of a measurement of the speed of light emitted from an orbiting body, at the phases in its orbit when it is moving directly towards or away from us. The orbiting body can be a particle in an accelerator, or a binary star orbiting its companion, but in either case the measured light speed is determined to be independent of the speed of the emitter, to impressively high accuracy.

For further discussion of this experimental support, and references to further reading, see Appendix B, French (1968, chs. 2 and 3), and Barton (1999, §3.4).                                    [Exercise 2.6]

## 2.2.1 Synchronising Clocks

The second axiom states that the speed of light is a universal constant, and it is this that allows us to define a procedure for synchronising clocks.

Suppose that you and I are some distance apart from one another, and relatively at rest. We start off with two clocks which are going at the same rate, but which are not initially synchronised. I send a flash of light in your direction, and note the time when I do so, $t_1$. You hold up a mirror to reflect the light back at me, and also note the time at which the flash arrived, $t_{mid}^{you}$. I note the time of the returned flash, $t_2$.

From the round-trip time, and knowing the speed of light, I can work out how far you are from me (it's simply $c(t_2 - t_1)/2$). Since the light moved at the same speed in both directions, I know that the time of the reflection was $t_{mid}^{me} = (t_1 + t_2)/2$, and I can communicate this time to you in some convenient way. You will compare that time to your $t_{mid}^{you}$, and adjust the zero of your clock to match. From that point on, our clocks are synchronised.

I describe this procedure in detail, not because it is complicated or particularly surprising, but to illustrate the extent to which we are here proceeding extremely cautiously from the axioms, careful to avoid smuggling in any extra assumptions. If we accept the axioms as true of our world, then we are forced to accept SR.

> There are a few further subtleties to this procedure which are not important for our purposes; both Rindler (2006, §2.6) and Taylor & Wheeler (1992, §2.6), for example, give details. One subtlety is that you may notice that the above procedure assumes that the speed of light is the same in both directions: although I doubt anyone seriously thinks that the speed of light is different, out and back, it is surprisingly difficult (indeed, impossible to date) to devise a procedure which avoids this assuption, or which is capable of measuring the 'one-way' speed of light.

## Exercises

### Exercise 2.1 (§2.1)

Consider a rocket at rest ($v_0 = 0$) at the origin of a frame $S$. At time $t = 0$, it starts to fire its rockets so that it moves along the $x$-axis, and at time $t = t_1$ we find the rocket moving at speed $v = v_1$. Consider a second frame $S'$, moving at speed $V$ along the $x$-axis, such that frames $S$ and $S'$ are in standard configuration (so that $x = x' = 0$ when $t = t' = 0$; the rocket of course has speed $v'$ and $v_1'$ in frame $S'$).

Presume that both the rocket and the primed frame are moving slowly enough that we can reasonably use the galilean transformation.

Work out the momentum of the rocket at $t = 0$ and $t = t_1$ in the two frames (that is, work out $p_0 = mv_0$, $p_1 = mv_1$, $p'_0 = mv'_0$ and $p'_1 = mv'_1$). Is momentum frame-invariant?

Work out the *change* in momentum in the two frames: is this frame-invariant?

Work out the kinetic energy and the change in kinetic energy in the two frames: are these frame-invariant?

## Exercise 2.2 (§2.1)

Suppose I have a fancy new cosmological theory that says that there's a special point in the universe – say, half-way between here and Andromeda – which is such that the gravitational constant $G$ changes depending on how far away from that point you are. Does this theory have a chance of being right? Why?                                                                    [$d^-$]

## Exercise 2.3 (§2.1)

From the constant-acceleration equations, you learned, early in your education, how to analyse projectile motion: you discovered that, for a projectile launched at speed $u$ at an angle $\theta$ to the horizontal, the time aloft was $t = 2u \sin\theta/g$ and the range was $x = u^2 \sin 2\theta/g$ (just to be clear, in this exercise we're still talking about galilean relativity, and the galilean transformation – we still haven't got to Special Relativity).

(i) Consider a train moving through a station at a speed $U$. Draw a diagram of this situation, indicating the two reference frames $S$ and $S'$ in standard configuration.

(ii) Someone on the train (frame $S'$) throws a ball vertically into the air at a speed $u'$ (i.e., $u'_x = 0$ and $u'_y = u'$). Calculate its time aloft, $t'$, and range, $x'$, measured in the moving frame, and reassure yourself that the latter is zero (i.e., the ball comes vertically downwards and the person catches it again).

(iii) Do the same calculation from the point of view of a person standing on the platform (i.e., work out the initial $u_x$ and $u_y$ for the ball, and work out the consequent projectile motion).

(iv) Finally, take your answer from part (ii), and use the GT in Eq. (2.1) to substitute the primed quantities. Confirm that you get the same answers as in part (iii).

### Exercise 2.4 (§2.1.1)

Consider the wave equation

$$\frac{\partial^2 \phi}{\partial x^2} + \frac{\partial^2 \phi}{\partial y^2} + \frac{\partial^2 \phi}{\partial z^2} - \frac{1}{c^2}\frac{\partial^2 \phi}{\partial t^2} = 0. \tag{i}$$

Take

$$\frac{\partial}{\partial x} = \frac{\partial x'}{\partial x}\frac{\partial}{\partial x'} + \frac{\partial t'}{\partial x}\frac{\partial}{\partial t'}$$

$$\frac{\partial}{\partial t} = \frac{\partial x'}{\partial t}\frac{\partial}{\partial x'} + \frac{\partial t'}{\partial t}\frac{\partial}{\partial t'}$$

$$\frac{\partial}{\partial y} = \frac{\partial}{\partial y'} \qquad \frac{\partial}{\partial z} = \frac{\partial}{\partial z'}.$$

Using the GT, Eq. (2.1), show that Eq. (i) does *not* transform into the same form under a GT.                              $[\,d^+\,]$

### Exercise 2.5 (§2.1.2)

I have a friend moving past me in a rocket at a relativistic speed, and I observe her watch to be ticking slower than mine (as we will discover later). She examines my watch as I do this: is it ticking faster or slower than hers?
$[\,d^-\,]$

### Exercise 2.6 (§2.2)

You are on a train moving through a station at $50\,\mathrm{m\,s}^{-1}$, and you throw a ball forwards at $10\,\mathrm{m\,s}^{-1}$: what is the speed of the ball as measured by someone on the platform?

Now you are on a train moving at half the speed of light, and you shine a torch forwards: what is the speed of the light from the torch as measured by someone on the platform?

1. $0.5c$
2. $c$
3. $1.5c$
4. $2c$                              $[\,d^-\,]$

# 3

# Length Contraction and Time Dilation

Before thir eyes in sudden view appear
The secrets of the hoarie deep, a dark
Illimitable Ocean without bound,
Without dimension, where length, breadth, & highth,
And time and place are lost

*John Milton,* Paradise Lost, *II, 890–894*

In this chapter we will examine some immediate consequences of the axioms
of Chapter 2, and develop some qualitative understanding of these before
we mathematise things in the next chapter.

*Aims*: you should:

3.1. understand why the concept of simultaneity is problematic in the context of SR, and how we resolve these problems.

## 3.1 Simultaneity

Imagine standing in the centre of a train carriage,[1] with suitably agile friends
at either end: Fred (at the Front) and Barbara (at the Back). At a prearranged
time, say time '0' on your carefully synchronised watches, you fire off a
flashbulb and your friends note down the time showing on their watches
when the flash reaches them (Figure 3.1). Since you are standing in the

---

[1] The argument below ultimately originates from Einstein's popular book about
relativity (Einstein 1920), first published in English in 1920. It is clearly ancestral to the
multiple versions, involving planes, trains, automobiles and rockets, in both popular and
professional books on relativity. The particular variant described here is most immediately
descended from Rindler's version (2006).

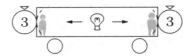

**Figure 3.1** A flashbulb in a stationary train.

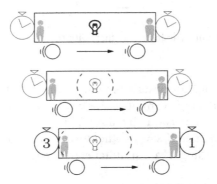

**Figure 3.2** Passing trains.

middle of the carriage, Fred's and Barbara's times must be the same as each other. Comparing notes afterwards, you all find that it took some time for the flash to travel from the middle of the carriage to the end, and that your friends have noted down the same arrival time on their watches, time '3', say. In other words, Fred's watch reading '3', and Barbara's watch reading '3', are *simultaneous* events in the frame of the carriage.

> ⚠ These watches are obviously not calibrated in seconds, but these could be sensible values if the watches are telling time in the natural relativistic time unit of light-metres. See Section 4.1 below.

Suppose now that this train is moving through a station as all this goes on, and you look from the platform into the carriage – what would you see from this point of view? You would see the light from the flash move both forwards towards Fred and backwards towards Barbara, but remember that you would *not* see the light moving forwards faster than the speed of light – its speed would not be enhanced by the motion of the train – nor would you see the light moving backwards at less than $c$. Since the back of the train is rushing towards where the light was emitted, the flash would naturally get to Barbara first, as illustrated in Figure 3.2. At that point Barbara's watch *must* read '3', since the flash meeting her and her watch reading '3' are simultaneous at the same point in space at exactly the same time, and so

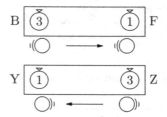

**Figure 3.3** Two trains passing each other, with the observers' locations indicated.

must be simultaneous for observers in any frame. But at this point, the light moving towards Fred cannot yet have caught up with him: since the light reaches Fred when *his* watch reads '3', his watch must still be reading something less than that, '1', say. In other words, Barbara's watch reading '3' and Fred's watch reading '1' are simultaneous events in *your* inertial frame on the platform.

What is going on here? Are these events simultaneous or not? What this tells us is that our notion of simultaneity is initially rather naïve, and that we have to be very careful exactly what we mean when we talk of events as being simultaneous. The only case where two events are quite unambiguously simultaneous is if they take place at exactly the same point in space.

## 3.2 Length Contraction and Time Dilation, Qualitatively

We're not finished with the trains, yet. Imagine now we're standing on the platform and see, this time, two trains go past in opposite directions, at the same speed as in the previous section. As well as Barbara and Fred, we have Yvette and Zebedee stationed at the very front and back of the lower carriage, also with clocks and well-sharpened pencils. We've cunningly arranged the speeds, timetable and flashbulbs so that we can get the set of observations represented in Figure 3.3, where the light has reached both rear observers and neither front one. This time, because the lower carriage is moving in the opposite direction, the same argument as in Section 3.1 has the clocks showing times '1' and '3' the opposite way around.

We are going to presume that the trains pass close enough to each other, and that the various observers have their noses pressed firmly enough against the windows, that observers in the two trains can be taken to be in the same

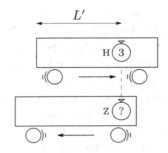

**Figure 3.4** Two trains, a little later.

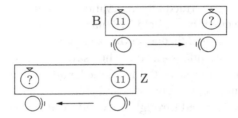

**Figure 3.5** The two trains, a little later still.

location, and thus can make 'local' observations of each other's clocks. In Figure 3.3, for example, Barbara can at this instant legitimately read the time on her watch *and* the time on Yvette's. She cannot read the time on Fred's watch, nor Yvette's watch at any other time, because those would not be 'local' observations.

Now pause a moment, and take another set of observations, shown in Figure 3.4 (we'll come back to this image in a moment). And then take a final set of observations when the two rear observers are beside each other, this time getting Figure 3.5.

> The images are intended to represent a suitable set of observers lined up along the platform edge. You can if you wish think of these two figures as 'photographs' of the scene, if that would be helpful. Note, however, that these 'photographs' are emphatically *not* what you would see if you used a real camera. A real such photo would include optical effects such as aberration and the Doppler effect, along with effects arising from the propagation time between different points in the scene, which would hopelessly confuse the issue here.

After this, the various observers calm down, amble together, and compare notes. Barbara (standing at the back of the top carriage) could remark 'I saw the front of the other carriage pass me when my clock was reading "3"' (this is perfectly correct, as you can confirm by looking at the positions in Figure 3.3). To which Fred would reply 'But Zebedee passed me when my clock showed time "1" – he must have been well past me at time "3"' (also true, from the same figure, and since Fred and Barbara's clocks are synchronised in their frame). Someone other than Fred was beside Zebedee at time '3'. From this they, and we, can quite correctly remark that the carriage they observed moving past them was measured to be shorter than their own. They have measured the length of a moving carriage using the procedure of Section 1.4, and found that it is shorter, by some amount which we can't calculate just yet, than a similar carriage (their own) which they can measure at leisure when stationary. This is *length contraction*.

In particular, this intermediate observation is what we see in Figure 3.4, where we see the clock being held by Hilary. Hilary is the observer who, at time '3', is adjacent to Zebedee, at the back of the lower carriage. Hilary and Barbara can conclude that, since they saw respectively the back and the front of the lower carriage at time '3', the length of that carriage in *their* frame is the distance between Barbara and Hilary in *their* frame, namely $L'$.

Fred then says 'I looked through the window at Zebedee's clock, and I noticed that it was reading "3", when mine was reading "1" – it was two units fast' (this also is true, as you can see from Figure 3.3). Barbara says 'Well, I *also* saw Zebedee's clock a bit later [in Figure 3.5], and it was reading "11", just like mine – it wasn't fast at all.' That is, there were $11 - 1 = 10$ units of time between Zebedee meeting Fred and Zebedee meeting Barbara, as measured on Fred and Barbara's clocks, but only $11 - 3 = 8$ units of time between those *same* two events, as recorded on Zebedee's watch.

Fred and Barbara know that their own clocks were synchronised throughout the encounter (they can make sure that their clocks are synchronised at some point, and they know that they both go at the same rate), so they can only conclude (correctly) that the clock they both saw was going more slowly than theirs were. Zebedee's clock has measured a smaller amount of time passing, between encountering Fred and encountering Barbara, than has been measured to pass in Fred and Barbara's frame. If we say 'time in the other carriage is passing more slowly', this is what we mean (note, by the way, that we have nothing to say about Yvette's clock since, in Figure 3.3 and Figure 3.5, it is only Barbara who is briefly adjacent to Yvette's clock, whereas both make successive local observations of Zebedee's).

The startling thing is that Yvette and Zebedee would come to precisely the same conclusions. Because this setup is perfectly symmetrical, they would measure Barbara's clock to be moving slowly, and Barbara and Fred's carriage to be length contracted. There is no sense in which one of the carriages is *absolutely* shorter than the other.

We are not at this point able to calculate how much the lengths of the carriages have contracted. We will come back to this shortly, in Section 3.4 (and see also Exercise 5.6).                                    [Exercises 3.1 & 3.2]

## 3.3 The Light Clock

Having persuaded ourselves (I hope) of the existence of time dilation and length contraction, it is easy to put numbers to the effects.

Imagine you're on a train which is, again, moving through a station at some non-relativistic speed. You throw a ball into the air with one hand, and then catch it with the other: how would you describe this? You'd say that the ball started off in your left hand, followed a parabolic path (like anything thrown), landed in your right hand, and that it took one second (say) to do it all.

Now imagine you're on the platform watching this go on: how would you describe it now? You'd say that it started off at the start of the platform, landed a good way down the platform, and took one second to do it. These two perceptions agree that the ball follows a parabolic path (different parabolae, yes, but parabolae nonetheless – this is the Relativity Principle at work), but they disagree on how far the ball travelled in flight. That disagreement is easily explained: from the point of view of the observer on the platform, the ball was travelling very quickly, since it had the train's speed as well as its own; so of *course* it covered more ground before it landed again.

None of this is mysterious. I've possibly made it sound mysterious by the elaborate way I've described it. But I've described it that way to pull this perfectly normal situation into line with the next step, the *light clock*.

The light clock (see Figure 3.6) is an idealised timekeeper, in which a flash of light leaves a bulb, bounces off a mirror, and returns – this is one 'tick' of the clock.

We set things up so that the clock's mirrors are arranged perpendicular to the clock's motion. This means that both the stationary and the moving observer measure the same separation between them – there is no length contraction perpendicular to the motion. To see that this must be so, consider the following *reductio ad absurdum*. Imagine there were a perpendicular length contraction. Then observers measuring a train moving along a parallel railway track would see the train getting narrower; specifically, the train's axles would get shorter, so that at some speed the train would be derailed with its wheels lying between the rails. However, from the point

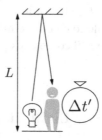

**Figure 3.6** Light clock.

of view of the observers in the train, it is the rails which would be moving; therefore, if there were a perpendicular length contraction, the measured distance between the rails would become shorter, and at some speed the train would be derailed with its wheels lying *outside* the rails. These statements must be either both true or both false; they contradict one another, so they cannot both be true, and they must therefore both be false; they are both consequences of the supposition that there exists a perpendicular length contraction; so that statement must in turn be false, demonstrating that there can be no such contraction.

If the mirror and the flashbulb are a distance $L$ apart, and I, standing by the light clock in Figure 3.6, time the round trip as $\Delta t'$ seconds, then, since the speed of light is the constant $c$,

$$2L = c\Delta t'. \tag{3.1}$$

Note that $\Delta t'$ here is the time interval on my watch, standing alongside and moving with the clock.

Now observe the light clock, stationary on the train going through the station, as you watch it from the platform (see Figure 3.7). The clock is in motion, at a speed $v$, so that the flash of light is reflected by the opposite mirror when it is a little way down the platform, and detected when it is still further on. Here, one tick is timed as $\Delta t$ seconds, during which time the clock will have moved a distance $v\Delta t$ down the platform.

How far has the light travelled? We know the light travelled at a speed $c$, from the second axiom, and we timed its round trip at $\Delta t$ seconds, so the light beam must have travelled a distance $c\Delta t$ in the time that the clock itself travelled a distance $v\Delta t$. But from the figure,

$$\left(\frac{c\Delta t}{2}\right)^2 = L^2 + \left(\frac{v\Delta t}{2}\right)^2. \tag{3.2}$$

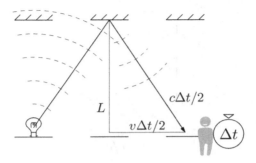

**Figure 3.7** Light clock: the observer is standing on the station platform, and times the round-trip time as $\Delta t$ on their clock.

Substituting $L$ from Eq. (3.1), and rearranging, we find

$$\Delta t' = \left(1 - \frac{v^2}{c^2}\right)^{1/2} \Delta t = \frac{\Delta t}{\gamma}, \tag{3.3}$$

where the factor $\gamma = \gamma(v)$ is defined as

$$\gamma = \left(1 - \frac{v^2}{c^2}\right)^{-1/2}. \tag{3.4}$$

Now, the important thing about Eq. (3.3) is that it involves $\Delta t'$, the time for the clock to 'tick' as measured by me, standing next to it on the train, and it involves $\Delta t$, the time as measured by you on the platform, and *they are not the same*.

How can this possibly be? Why is this different from the perfectly reasonable behaviour of the ball thrown down the carriage, in the non-relativistic example at the beginning of this section? The difference is that when you watched the ball from the platform, you saw it move with the speed it was given plus the speed of the train – in other words, the person on the platform and the person on the train had a perfectly reasonable disagreement about the speed of the ball, which resulted in them agreeing on the time the ball was in flight. In the relativistic example, however, both of them *agree* on the speed of the light in the light clock, as the second axiom says they must. Something has to give, and the result is that the two observers disagree on how long the light takes for a circuit.

So at least one of the clocks is broken? They're both in perfect working order (this is again the 'clock hypothesis' which was briefly mentioned in Section 1.6). They only work properly when they're stationary? No, the

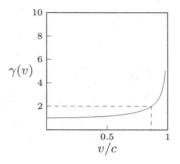

**Figure 3.8** The gamma function, or Lorentz factor, $\gamma = (1 - v^2/c^2)^{-1/2}$, plotted as a function of $v/c$.

Relativity Principle tells us that there's no sense in which either of them is 'more stationary' than the other, so that the clocks work in exactly the same way whether they're moving or not. No...

> *Both clocks are perfectly accurately measuring the passage of time; time is flowing differently for the two observers.*

Think back to the 'taffrail log' of Section 1.6: the clocks carried by the moving and by the stationary observer have been dragged through different distances in time.

The factor $\gamma = (1 - v^2/c^2)^{-1/2}$ is known by a couple of different names, including the 'Lorentz factor'; you will become very familiar with this expression. As you can see from Eq. (3.3) or Eq. (3.5), it indicates how significant the relativistic effects are, for a given velocity $v$. At $v = 0$, it has the value $\gamma(0) = 1$, showing that there is no time dilation for a stationary frame. As you can see in Figure 3.8, the factor stays very close to one for much of its range, and even at nearly 90% of the speed of light ($v/c = \sqrt{3}/2 = 0.87$) it is still only $\gamma = 2$; as $v$ increases beyond this, however, the Lorentz factor grows very rapidly, becoming infinite at $v = c$. Thus there is no point at which relativistic effects suddenly switch on – we are always in a relativistic universe – but they are ignorable at lower speeds.          [Exercises 3.3 & 3.4]

## 3.4 The Horizontal Light Clock: Length Contraction

In Section 3.3 we saw how to use the light clock to obtain the time-dilation formula, Eq. (3.3). Just for an encore, and as another illustration of how we can do calculations using just the ideas we have so far, let's consider a variant of this light clock, which lets us derive the length-contraction formula (we obtain this result again, in a rather more compact way, in Section 5.5.1, and that version may be a little easier to follow).

Consider a *horizontal* light clock, such as a train carriage of length $L_0$ with a mirror at the front. Take the carriage to be at rest in frame $S'$, with the rear of the carriage at the origin of $S'$. The carriage is measured to be length $L$ in the frame of the station; we know from Section 3.2 that the distance $L$ will be less than $L_0$, as a result of length contraction, but we do not know, at this point, just how much shorter it will be. This length, $L_0$ (which we could write $L'$ if we preferred), is sometimes referred to as the *proper length* of the carriage, meaning its length in a frame in which it is stationary.

Imagine a light flash at $x' = 0$ at time $t' = 0$, which is reflected from the mirror at the front, which we will call event ①, at times denoted $t_1$ and $t_1'$ in the two frames, and detected at the back of the carriage again, in event ②, at times $t_2$ and $t_2'$. By considering the round-trip time in the two frames, we can find expressions for $L$ and $L_0$ in terms of $v$ and $c$ (you might want to try deriving this yourself, before reading on).

In frame $S'$, the analysis is simple. The light travels a distance $2L_0$ in time $t_2'$, so

$$2L_0 = ct_2'.$$

Because the original flash and event ② happen at the *same location* in $S'$ – we are reading the same clock twice – we can use the time-dilation formula, Eq. (3.3), to relate $t_2$ and $t_2'$, and discover that $t_2 = \gamma t_2'$ (the time-dilation formula relates time *intervals*, not time coordinates, so we are here relating the intervals between time $t = t' = 0$, when the light flashed, and times $t_2$ and $t_2'$ when the reflection returned).

In frame $S$, we examine the light's travel from the origin to event ①, and separately its travel between events ① and ②. The light travels at $c$, so $x_1 = ct_1$. In doing so, it has travelled the length of the carriage (as measured in frame $S$), plus the distance the carriage has travelled in that time, so that $x_1 = L + vt_1$. Thus

$$t_1 = \frac{L}{c - v}.$$

On the return leg, similarly, we find $x_1 - x_2 = c(t_2 - t_1)$ and $x_1 - x_2 =$

$L - v(t_2 - t_1)$. Adding $t_1 + (t_2 - t_1)$ we find

$$t_2 = \frac{L}{c-v} + \frac{L}{c+v} = \frac{2Lc}{c^2 - v^2} = \frac{2L}{c(1 - v^2/c^2)} = \gamma^2 \frac{2L}{c}$$

$$= \gamma t_2' = \gamma \frac{2L_0}{c},$$

and thus, after a little rearrangement, that

$$L = \frac{L_0}{\gamma}. \tag{3.5}$$

The length of the carriage in the platform frame, $L$, is less than its length in the carriage frame.

I've added a few extra remarks about length contraction and time dilation in Appendix D.                                         [Exercise 3.5]

## 3.5 Is There Anything I Can Hold on To?

At this point you may be feeling rather seasick. People tend to find relativity rather disorienting, as more and more pillars of their dynamical intuition are kicked away. You can end up in the situation where you trust none of your steps in any direction, and find yourself unable to move at all, for fear that the whole edifice will come tumbling down.

Distinguish between what you know, and what you intuit. In fact, rather little of what you know has changed: it comes down to not much more than the relativity of simultaneity ('not everyone agrees that two events are simultaneous'), which leads to length contraction and time dilation as fairly direct consequences of the second axiom. Unfortunately, both of these eat away at our intuition of how moving objects behave.

Look at what has *not* changed, however. For example, in our discussion of the trains in Figure 3.2, it is still true that the backwards-moving light flash hit the rear of the carriage before the forwards-moving one hit the front because, reasonably enough, the carriage rear was moving *into* the flash. It is still true that the order of events at a single point in space is absolutely fixed. It is still true that when Fred and Barbara looked at their own watches those watches told them the local time accurately, and when they looked at the nearby watches on the other train (which are, in principle, at the *same* point in space and time as Fred and Barbara), they could reasonably measure, or observe, what they saw there.

The focus on local measurements can help us understand what is going on. If we concentrate on working out what the participants would measure

– that is, on what they would see happening at their own point in space and time – rather than on what we intuitively expect them to see, then when we find ourselves puzzled, saying 'that couldn't possibly happen, because…', we have someplace to start.[2]

## Exercises

### Exercise 3.1 (§3.2)

Which of the following statements are true, referring to the observers at the front and back of the train carriage, in the discussion of Section 3.2?

1. Fred and Barbara's watches stay synchronised with each other.
2. Fred and Barbara's watches stay synchronised with the clocks in the other carriage.
3. Fred and Barbara measure their carriage to get shorter when they're moving.                                                          [ $d^-$ ]

### Exercise 3.2 (§3.2)

Which of the following statements are true, referring to the observers at the front and back of the train carriage, in specifically Figure 3.2?

1. Fred and Barbara measure the other carriage to get shorter when it's moving relative to them.
2. Fred and Barbara measure the speed of light in the other carriage to be less than $c$.
3. By measuring the Doppler shift of a light signal sent from the back of the carriage to the front, the two observers can determine the carriage's velocity to any desired accuracy.

---

[2] In this focus on (local) measurement, we can see the influence of the philosophical 'positivism' which influenced Einstein in his development of SR. This is the claim that it is observations of the external world that give us knowledge, rather than *a priori* suppositions. There is some apparent tension with the rather axiomatic approach to SR which Einstein then develops, but while the approach is rather abstract, the contact between the theory and the world is via tangible events or measurements, rather than abstractions such as 'absolute space'.

## Exercise 3.3 (§3.3)

When discussing the light clock, just before Eq. (3.2), we we saw the phrase 'one tick is timed as $\Delta t$ seconds'. This is $\Delta t$ on whose clock?

1. The watch of a person on the train.
2. The watch of a person on the platform edge.
3. The station clock.
4. The time attached to the photon of light.                    $[\,d^-\,]$

## Exercise 3.4 (§3.3)

Repeat the analysis of Section 3.3 for a 'tennis-ball clock'. Balls are fired horizontally across a carriage (moving at a non-relativistic speed): the time it takes them to bounce back to the starting point is one 'tick' of the clock. What is the crucial point at which the two analyses diverge?

## Exercise 3.5 (§3.4)

Consider the following argument.

Consider two frames $S$ and $S'$ in standard configuration; there is a rod of length $L_0$ laid along the $x'$-axis with one end at the origin. What is the length of this rod, $L$, as measured in the $S$ frame?

Imagine our line of observers stretching along the $x$-axis, all with synchronised clocks (showing $t$). At time $t = 0$, when the origins of the two frames coincide (so that one end of the rod is at coordinate $x = 0$), precisely one of the observers in the $S$ frame will be standing opposite the far end of the rod, located at coordinate $x' = L_0$ in the $S'$ frame; this observer has coordinate $x = L$.

Let us arrange for a firework to go off at the origin at time $t = t' = 0$: travelling at speed $c$, the light from this will reach our observers at coordinates $x = L$ and $x' = L_0$ at times $t = L/c$ and $t' = L_0/c$. But, from Eq. (3.3), we know that any two events separated by time $t'$ in the moving frame would be measured to be separated by time $t = \gamma t'$ in the stationary one. Thus, the time intervals $L/c$ and $L_0/c$ are related, and so

$$L = \gamma L_0 \qquad \text{[WRONG!]}.$$

This conclusion is *wrong*, as comparison with Eq. (3.5) shows, so the argument is fallacious. What's wrong with it?                    $[\,d^+\,]$

# 4

# Spacetime and Geometry

But in the dynamic space of the living Rocket, the double integral has a different meaning. To integrate here is to operate on a rate of change so that time falls away: change is stilled.... 'Meters per second' will integrate to 'meters.' The moving vehicle is frozen, in space, to become architecture, and timeless. It was never launched. It will never fall.

*Thomas Pynchon,* Gravity's Rainbow

Chapter 3 has shown us that the axioms of SR take us into unexpected territory. Now it is time to explore the new landscape, and discover that geometry can help orient us.

*Aims*: you should:

4.1. appreciate the role of geometry in understanding spacetime, specifically the importance of the invariant interval and the Minkowski diagram; and
4.2. internalise the utility of units where $c = 1$, as the natural units for discussing events in spacetime.

In Chapter 3 we saw that if a set of observers, in motion relative to each other, separately observe a common pair of events, they will make separate measurements of the events' coordinates in space and time, and produce *different* values for those positions. This is not simply due to insignificant differences such as having differing origins – having the observers' frames in standard configuration (Section 1.7) deals with that. Nor is it due to using different measurement scales – numbers can be trivially converted from one unit to another, and in any case we are about to discuss (Section 4.1) the way we systematically handle this problem. It is also *not* due to any effect of light-travel time: all the observers make measurements only of events immediately local to them, and the observations we have of 'distant' events are relayed from observers local, in space and time, to those events. Instead, we have discovered that not only will different observers disagree about the order in which separated events are observed to occur (Section 3.1), but (partly as a consequence, and as we saw in Section 3.2), observers in different frames will not agree about the separation in time, or the separation in space, between two events. Such separations are things we would think are solidly established (once we have put aside the trivial complications we have just discussed); it is disturbing to find that they are not.

The relationship between the different measurements is not random – the coordinates obtained by one observer are systematically related to those obtained by another. You are already familiar with this general idea: the space between objects in your environment is not changed when we change the units we use to measure it (such as kilometres versus inches), and there are all sorts of regularities in the network of such distances, such as Pythagoras's theorem, or that the internal angles of a triangle add up to 180°. That is, we are familiar with the idea of *geometry*. We also learn to become comfortable with the idea that the rules of geometry in our immediate environment (Pythagoras, and all that) have to be adjusted when we consider Earth-sized distances, or distances on the sky: the geometry of the surface of a sphere has different rules from those that Euclid wrote down. We are about to learn a further set of geometrical rules, and learn that we can make sense of the things we encountered in the last chapter by seeing them as the consequences of a set of geometrical rules that we have not known about before now.

This chapter takes a first look at these ideas, before I pull them together in a geometrical approach to the *Lorentz transformation* (LT, Eq. (5.6)), in the next chapter. This way of approaching Special Relativity is, I think, tremendously powerful, and (a separate advantage) creates a natural bridge to General Relativity, which we will explore a little more in Appendix A.

Before we can properly embark on this, however, we must look more carefully at the way in which time and space are interrelated.

And before we do that, it is convenient to get our units of measurement straight.

## 4.1 Natural Units

We tend to choose units appropriate to our topic of conversation – hours for appointments, days for calendars, metres for our height, kilometres for our maps, and parsecs for galaxies. What units should we use when we are talking about moving near the speed of light? The fact that the speed of light is $3 \times 10^8 \, \text{m s}^{-1}$ tells us that if we persist in using metres and seconds, we'll get a lot of very large or very small numbers, and lose all hope of developing much in the way of intuition.

If the problem were merely one of large numbers, we could settle on the gigametre as our usual unit, give it a handy name, and be done with the question. However we will soon discover that space and time are not as distinct as might at first appear, and that having different names for the separations in these directions can obscure this. It's as if we had decided to measure distances east–west in kilometres and distances north–south in inches. Much better is to measure the two in the *same* units.

One possibility is to use time as a measure of distance. We do this naturally when we talk of the Earth being about 8 light-minutes from the Sun, or the nearest star being a little more than 4 light-years away. We can also talk of the light-second, of $1 \, \text{s} = 3 \times 10^8 \, \text{m}$ (the Sun has a diameter of 4.6 s), or the light-nanosecond of 30 cm. In these units, light moves at a speed of one light-second per second, $c = 1 \, \text{s s}^{-1}$, or one light-year per year, $c = 1 \, \text{yr yr}^{-1}$; that is, $c = 1$, a unitless number.[1]

There's nothing wrong with this in principle, but a more common convention in this context is to instead use space as a measure of time, and use the *light-metre* as our time unit, with a light-metre being the *time* it takes for light to travel one metre; that is, a little more than 3.3 ns. In these units, light travels a distance of one metre in a time of one light-metre, so again $c = 1 \, \text{m m}^{-1}$ as a unitless number (we did this, in fact, in Section 3.1). These are referred to as *natural units*. Talking of 'metres of time' feels initially unnatural, but it's not any odder than talking about (light-)years as a distance, and we're generally quite comfortable with that. When we need to distinguish them, we refer to SI units – that is, metres and seconds – as *physical units*.

In fact, since 1983, the International Standard definition of the metre is that it is the distance light travels in $1/299\,792\,458$ seconds; that is, the speed of light is $299\,792\,458 \, \text{m s}^{-1}$ *by definition*, without measurement uncertainty, and so $c$ is therefore demoted to being merely a conversion factor between two different units of time. The 'second' being used here is the SI second, which is defined so that a particular atomic transition in caesium has a specific defined frequency in hertz (for further documentation of the various resolutions here, see the 'SI Brochure' (BIPM 2019)). International Atomic

---

[1] It is not *dimensionless*, since it still has the dimensions $LT^{-1}$, but since both dimensions use the same units, they cancel.

Time (TAI) consists of a count of such seconds from a particular epoch in 1977.

In *exactly* the same sense, the inch is *defined* as 25.4 mm, and this figure of 25.4 is merely a conversion factor between two different, and only historically distinct, units of length.[2] It does look odd to see the unit conversion written as $1 = 3 \times 10^8$ m s$^{-1}$; it would be similarly odd, but similarly rational, to write the definition of the inch as $1 = 25.4$ mm in$^{-1}$.

It is the fact that the speed of light is a universal frame-independent constant – one of the axioms of SR – that allows us to use this speed as a conversion factor in this way. This procedure would make no sense in newtonian physics, where Eq. (2.1b) asserts that the speed of light is a frame-dependent quantity.

There are several advantages to all this. (i) As we will learn below, space and time are not as distinct as our intuition might suggest, but having different units for the two 'directions' can obscure this; and (ii) if we measure time in light-metres, then we no longer need the conversion factor $c$ in our equations, which are consequently simpler (from here on, we'll refer to the time units as simply 'metres', and I'll let you mentally insert the 'light-' prefix yourself). We also quote other speeds in these units of metres per metre, so that all speeds are unitless and less than one. For the rest of this text, unless otherwise noted, **all dimensions and speeds will be quoted in natural units:** $c = 1$. Most relativity textbooks do not do this, and quote equations in physical units instead.[3] You can find a table of these later on, in Section 5.6.

It is easy, once you have a little practice, to convert values and equations between the different systems of units (or at least, it is no more confusing than unit conversions normally are).

For example, to convert $10 \, \text{J} = 10 \, \text{kg} \, \text{m}^2 \, \text{s}^{-2}$ to natural units, we could proceed in two ways. Since $c = 1$, we have $1 \, \text{s} = 3 \times 10^8 \, \text{m}$, and so

---

[2] If there really were some convention that east–west distances were in kilometres and north–south ones in inches, then land surveyors would all be familiar with the conversion factor $k = 2.54 \times 10^{-5}$ km in$^{-1}$, and the equations in their textbooks would be littered with factors of $k$ and $k^2$ and so on. Instead, surveyors would surely write their textbook equations in 'natural units' where $k = 1$, and when they want to calculate 'real-world' numbers, they would have to learn the techniques for either converting those measurements to natural units, or re-inserting the 'missing' factors of $k$ into their equations, to re-obtain 'physical units'. But *of course* land surveyors would amongst themselves avoid this km/in conversion; and *of course* we as spacetime-surveyors want to avoid the analogous convention of heterogeneous units.

[3] Natural units are not a modern innovation, cruelly designed to confuse students. Eddington used natural units throughout his 'Report' (1920), which was the first book-length description in English of Special and General Relativity.

$1\,\mathrm{s}^{-2} = (9 \times 10^{16})^{-1}\,\mathrm{m}^{-2}$. So $10\,\mathrm{kg}\,\mathrm{m}^2\,\mathrm{s}^{-2} = 10\,\mathrm{kg}\,\mathrm{m}^2 \times (9 \times 10^{16})^{-1}\,\mathrm{m}^{-2} = 1.1 \times 10^{-16}\,\mathrm{kg}$.

Alternatively (saying the same thing in a slightly different way), we can write $1 = 3 \times 10^8\,\mathrm{m}\,\mathrm{s}^{-1}$, or $1 = (3 \times 10^8)^{-1}\,\mathrm{s}\,\mathrm{m}^{-1}$. Thus

$$\begin{aligned} 10\,\mathrm{J} &= 10\,\mathrm{kg}\,\mathrm{m}^2\,\mathrm{s}^{-2} \times (1)^2 \\ &= 10\,\mathrm{kg}\,\mathrm{m}^2\,\mathrm{s}^{-2} \times (3 \times 10^8)^{-2}\,\mathrm{s}^2\,\mathrm{m}^{-2} \\ &= 1.1 \times 10^{-16}\,\mathrm{kg}. \end{aligned}$$

To convert from natural units to physical ones, it is useful to consider the dimensions of the quantities in question. Thus to convert an energy $1.1 \times 10^{-16}\,\mathrm{kg}$ to physical units, we recall that the dimensions of energy are $\mathrm{ML}^2\mathrm{T}^{-2}$, and thus write

$$\begin{aligned} E &= 1.1 \times 10^{-16}\,\mathrm{kg} \\ &= 1.1 \times 10^{-16}\,\mathrm{kg}\,\mathrm{m}^2\,\mathrm{m}^{-2}, \quad \text{recalling our time units are m} \\ &= 1.1 \times 10^{-16}\,\mathrm{kg}\,\mathrm{m}^2 \times (3 \times 10^8\,\mathrm{s}^{-1})^2, \quad \text{unit conversion} \\ &= 10\,\mathrm{kg}\,\mathrm{m}^2\,\mathrm{s}^{-2} = 10\,\mathrm{J}. \end{aligned}$$

It helps to recall that, with $1\,\mathrm{m} = (3 \times 10^8)^{-1}\,\mathrm{s}$, a metre is a very small unit of time, or that with $1\,\mathrm{m}^{-1} = 3 \times 10^8\,\mathrm{Hz}$ something that repeats once per metre-of-time is happening at a very high frequency. Similarly, it might be useful to think that when we write an energy as $E = 1\,\mathrm{kg}$, what we *mean* is $E = 1\,\mathrm{kg}\,\mathrm{m}^2\,\mathrm{m}^{-2}$, to be dimensionally correct, but we've skipped writing the cancelling units.

It's also possible to write equations with and without explicit factors of $c$. We have seen the Lorentz factor $\gamma = (1 - v^2/c^2)^{-1/2}$. In units where $c = 1$, this simplifies to $\gamma = (1 - v^2)^{-1/2}$. This expression is dimensionally consistent because, in these units, $v$ has no units (note: it still has the dimensions of $\mathrm{LT}^{-1}$, but our choice of time units means that these cancel). To convert this expression back to units where $v$ and $c$ have physical units, we simply have to add enough factors of $c$ inside the expression to make it dimensionally consistent. Another way of thinking about this is to take the various factors of $c$ to be still present in the equations, but 'invisible' because they have numerical value $c = 1\,\mathrm{m}\,\mathrm{m}^{-1}$. Resurfacing these factors, by deducing what power of $c$ must be present, then allows us to put in the physical-units value of $c$.

In General Relativity, people tend to work in units where the gravitational constant $G$ is also 1, so that mass has the same units as distance and time; the effect is that the mass of an object has the same numerical value

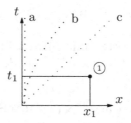

**Figure 4.1** A Minkowski diagram: this shows the sequence of events corresponding to a flashing bulb which is (a) at rest at $x = 0$, (b) accelerating from rest, and (c) moving at (nearly) the speed of light.

as the radius of the black hole that would result if the object were sufficiently compressed (in these terms, the Sun has a mass of 3 km). In relativistic quantum mechanics and particle physics, likewise, units are chosen so that Planck's constant $\hbar = c = 1$, and everything is quoted in energy units (usually electron-volts; see Section 7.5). As of 2018, all of the base units of SI are defined via conversion factors, in the same way as the metre – for example, the kilogramme is defined so that Planck's constant has a specific exact value (BIPM 2019).

[Exercises 4.1–4.3]

## 4.2 The Minkowski Diagram

How can we visualise motion?

Any event (say event ①) has a set of four coordinates $(t_1, x_1, y_1, z_1)$ in frame $S$. Since we will almost always restrict our attention to frames in standard configuration, and since there is no transverse contraction, the $y$ and $z$ coordinates are uninteresting. We can plot the remaining two coordinates $(t_1, x_1)$ on a diagram, to obtain something like Figure 4.1.

Imagine a flashing bulb which is stationary in frame $S$, at the origin. Each flash is a separate event: each event will have $x = 0$ and successively increasing $t$. We can plot these points to obtain the line marked 'a' in Figure 4.1; the line connecting these is the *worldline* of the bulb. If instead the bulb accelerates along the $x$-axis, then we obtain the line 'b'. As the bulb moves faster, the distance in space between two successive flashes, $\Delta x$, and their separation in time, $\Delta t$, are related as $\Delta x = v\Delta t$, giving the worldline's gradient on the diagram as $\Delta t / \Delta x = 1/v$. Since $v < 1$ for any physical object, an object's worldline always has a gradient larger than 45°.

If an object moves at speed $c$, then in 1 m of time it will travel 1 m along the $x$-axis, so its worldline will be a line at 45° to the axes, giving line 'c'.

**Figure 4.2** A flashbulb in the middle of a train carriage.

Notice that the worldline (b) approaches a 45° angle.

This diagram, characteristically with the $x$-axis horizontal and the $t$-axis vertical, is a *Minkowski diagram*, and the space it describes is referred to as *Minkowski space* or Minkowski spacetime.[4]

Consider now a train moving along at speed $v$, with a flashbulb in the centre, and a mirror at each end, as illustrated in Figure 4.2. Call the train's frame $S'$ (so positions in this frame will be described using coordinates $x'$ and $t'$), and locate its spatial origin (that is, coordinate $x'_1 = 0$) at the flashbulb; take the carriage to be 6 m long. We can identify a number of *events* which take place here:

- The bulb goes off at time $t'_1 = -3$ m; call this event ①.
- Events ② and ③ are that flash being reflected from mirrors on the $x'$-axis, at positions $x'_2 = +3$ m and $x'_3 = -3$ m (these reflections will happen simultaneously in frame $S'$, at $t'_2 = t'_3 = 0$).
- Event ④ is these reflected flashes reaching the centre again, at $t'_4 = +3$ m and $x'_4 = 0$ (note that this discussion has taken place entirely within frame $S'$; all these times are obtained simply from 'distance is speed times time').

The worldline of the flashbulb is simple, and is a straight line lying along the $t'$-axis, like worldline 'a' in Figure 4.1. We can plot these events, the worldline of the flashbulb, and the worldlines of the light flashes, on a Minkowski diagram, to obtain Figure 4.3(a). What does this sequence of events look like on a Minkowski diagram for frame $S$?

- Firstly, the worldline of the moving flashbulb is a slanted line in this frame; we know from the previous paragraph that this worldline lies along the $t'$-axis, so we can draw that axis in immediately. From $x = vt$, or rather $t = x/v$, we can see that this line has gradient $1/v$.

---

[4] Hermann Minkowski (1864–1909) introduced the idea in a talk in 1908 (see Section 5.3), where he emphasised the importance and fertility of this geometrical approach. The diagram is less exotic than it may at first appear: if it were drawn with $x$ vertical and $t$ as the horizontal axis, there would be nothing odd about it at all.

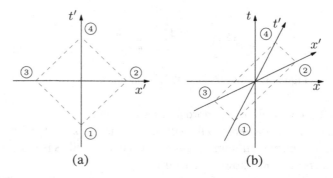

**Figure 4.3** A light flash, at event ①, reflected at events ② and ③, and re-observed at ④, as shown (a) in frame $S'$, in which the flashbulb is stationary, and (b) in $S$, in which the flashbulb is moving.

- The events ① and ④ happen at the location of the flashbulb, therefore they are located on the worldline of the flashbulb an equal distance either side of the origin. We'll shortly be able to work out just where on this worldline the events appear.
- So where are the events ② and ③? The light which travels from event ① to events ② and ③ must have a worldline which is angled at 45° (the fact that this is true *in all frames* is another statement of the second axiom); and the reflected light which travels from events ② and ③ back to event ④ must also be angled at 45°.
- The point where these light worldlines 'turn around' is the point where they are reflected, that is, at events ② and ③. We therefore see that the events ② and ③, which are simultaneous in the frame $S'$ (they have the same $t'$-coordinate), *are not simultaneous in the frame S*; this is yet another illustration of the relativity of simultaneity discussed in Section 3.1).
- The $t'$-axis joins the events ① and ④ in Figure 4.3, and the $x'$-axis joins events ② and ③. This must also be true for the newly discovered locations of these events in the $S$ frame.

This means that we can show, in Figure 4.3, the positions of the four events ① to ④ as they would be measured in frame $S$, as well as the positions of the axes of the $S'$ frame (the $t'$-axis is the worldline of the moving flashbulb, and the $x'$-axis is the line of simultaneity in the moving frame).

In Figure 4.3(a), events which happen at the same time in frame $S'$ – that is, which have the same $t'$-coordinate – are connected by a line parallel to the $x'$-axis. This remains true in Figure 4.3(b). Therefore in any Minkowski diagram, we can indicate the coordinates of a marked event, in each of

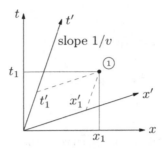

**Figure 4.4** The coordinates of an event projected onto the $(t, x)$ and $(t', x')$ axes of a Minkowski diagram.

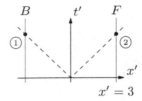

**Figure 4.5** The Minkowski diagram of the flashes in Figure 3.1, shown in the frame of the carriage.

the frames displayed in the diagram. We can see this in Figure 4.4, where the event ① has coordinates $(t_1, x_1)$ in frame $S$, and the *same* event has coordinates $(t_1', x_1')$ in frame $S'$.                    [Exercise 4.4]

### 4.2.1 Example Minkowski Diagram: the Moving Trains

We can draw on the Minkowski diagram the motions of the trains in Figure 3.1 and Figure 3.2. Call the frame fixed to the train-carriage $S'$, with its origin at the centre of the carriage. In this frame, the worldlines of the back and front of the carriage (marked $B$ and $F$ respectively) are stationary, so they run vertically on the diagram; the light flash from the centre runs forwards and back at speed $c = 1$, so is diagonal. We can therefore deduce the location of the two events ①, 'the flash reaches the back of the carriage', and ②, 'the flash reaches the front of the carriage', as shown in Figure 4.5.

How does this appear if we draw the Minkowski diagram of frame $S$, the station platform that the carriage is moving past? In this frame, the front and back are moving from left to right, so their worldlines slope upwards to the right ($c = 1$ in both frames, remember). The light flashes move diagonally as

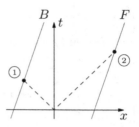

**Figure 4.6** The Minkowski diagram of the flashes in Figure 3.2.

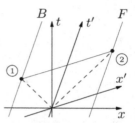

**Figure 4.7** The Minkowski diagram for Section 4.2.1, completed.

before. *The events are the same as before*: event ① is on both worldline $B$ and the worldline of the rearward-moving flash; event ② is on worldline $F$ and that of the forwards-moving flash. We thus obtain Figure 4.6, and we can immediately see that, although events ① and ② are simultaneous in $S'$ (in Figure 4.5), they are not simultaneous in frame $S$, as we saw in Section 3.1. We know that events ① and ② are simultaneous in $S'$, so that the $x'$-axis must be parallel to the line joining them, and that the worldlines $B$ and $F$ are parallel to the $t'$-axis, so we can go on to complete the Minkowski diagram in Figure 4.7 (it's unfortunately the case that Minkowski diagrams end up looking a lot more messy and confusing when complete, than they look when you're building them up by thinking through the description of a problem).

It may by now be clear that the geometry of the Minkowski diagram – the relationships between lengths and angles and what is and isn't perpendicular – is very different from the geometry we're used to. We need to find out more about that.

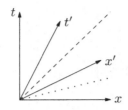

**Figure 4.8** The dots represent the events where the flashes from an offshore lighthouse beam are observed on a linear shoreline.

### 4.2.2 Example Minkowski Diagram: a Lighthouse

The Minkowski diagram can help us see what is going on in a given SR problem: if we plot the relevant events on the diagram, we can see their relationship more clearly. However, it can also support some arguments directly. For example (this argument is taken from Rindler (2006, §2.9)), imagine a flashing lighthouse beam being swept across a distant shore. If the shore is far enough away and the beam is turned quickly enough, the illuminated points can be made to travel arbitrarily fast – faster than the speed of light. We can plot such a path on the Minkowski diagram as the dotted line in Figure 4.8. Because it is travelling faster than light, it has a shallower gradient than the light beam at 45°. However, this means that there can be some frame in which the $x'$-axis lies along this dotted line – that is, some frame in which all the light flashes happen simultaneously. In any faster frame – the primed frame in Figure 4.8 – the illuminated points will be measured to travel in the opposite direction. This indicates that no causal influence – no light flash or anything travelling slower than it – can travel from one dot to the next.

Even when you cannot entirely solve a problem using a Minkowski diagram, they are extremely useful when you are trying to visualise a problem, prior, perhaps, to turning it into a set of events to manipulate using the Lorentz transformation we will learn about in the next chapter. Although it may seem a rather abstract way of noting down the information in a problem, with practice, you can often qualitatively solve the problem before plugging in any numbers at all.

## 4.3 Plane Rotations

Before the next step, we need a quick review of plane geometry – the geometry we're used to (that is, there's no Special Relativity in this section).

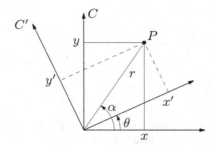

**Figure 4.9** Rotation in the plane: the point $P$ has different coordinates in two frames.

In Figure 4.9 we see frames $C$ and $C'$, where the latter has been rotated by an angle $\theta$ with respect to the former. The point $P$ has coordinates $(x, y)$ in the frame $C$ (here, as elsewhere, I am taking 'frame' and 'coordinate system' to be synonyms). The *same* point $P$ has coordinates $(x', y')$ in the frame $C'$; what is the relationship between the two sets of coordinates $(x, y)$ and $(x', y')$?

A little geometry (observe that the distance $r$ is the same in both frames, express $x$, $y$, $x'$ and $y'$ in terms of $\alpha$, $\theta$ and $r$, and eliminate $r$ and $\alpha$) gives

$$x' = \quad x\cos\theta + y\sin\theta \qquad\qquad (4.1a)$$
$$y' = -x\sin\theta + y\cos\theta. \qquad\qquad (4.1b)$$

The point $P$ is distance $r$ from the origin. Pythagoras's theorem tells us that the coordinates in $S$ are such that $r^2 = x^2 + y^2$; the transformation in Eq. (4.1) additionally implies that $r^2 = x'^2 + y'^2$ (this can be seen directly from the figure, and is consistent with the result of explicitly calculating $x'^2 + y'^2$ using Eq. (4.1)). That is, although the coordinates of the point $P$ in the two frames are different, the distance $r$ is the same in both frames – it is an *invariant of the transformation*:

$$x^2 + y^2 = r^2 = x'^2 + y'^2. \qquad\qquad (4.2)$$

## 4.4 The Invariant Interval

Let us look again at Figures 4.4 and 4.9, and place them side-by-side in Figure 4.10.

These are starting to look alike. In both cases, we have a single position $P$, or event ①, described by two sets of coordinates in two frames. These

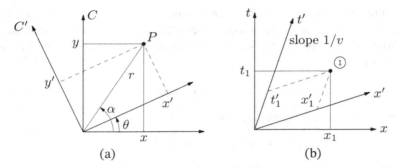

**Figure 4.10** (a) A length-preserving rotation in the euclidean plane (Figure 4.9); and (b) a length-preserving transformation in Minkowski space (Figure 4.4).

coordinates can be systematically related to each other. Given any pair of points $(x_1, y_1)$ and $(x_2, y_2)$, in the euclidean plane of Figure 4.10(a), we can define the separations $\Delta x = x_2 - x_1$ and $\Delta y = y_2 - y_1$, and use the constraint that the distance; note, by the way, that we read $\Delta x^2$ as $(\Delta x)^2$, and not $\Delta(x^2)$. The quantity $\Delta x^2 + \Delta y^2$ is an invariant under the transformation to derive Eq. (4.1). We can do the same for the distance $r^2 = x^2 + y^2 = x'^2 + y'^2$. Can we find a similar invariant for the transformation between frames which is represented by the Minkowski diagram, Figure 4.10(b)?

Look back at the light clock, illustrated in Figure 3.6 on p. 38: how 'far apart' are event ①, the light flashing, and event ②, the reflected light being received? In between these two events, the light has travelled a total distance 2L, so the temporal separation between events ① and ② is $\Delta t' = 2L$ (in units where the speed is $c = 1$), even though the spatial separation between the two events, as measured in this frame, is $\Delta x' = 0$. Inspired by the pythagorean distance above, let's write that separation-squared as $s'^2 = \Delta t'^2 = (2L)^2$. That interval involves only the time between the two events, but the two events happened at the same location $(\Delta x' = 0)$, so any contribution from that spatial separation will not be present in this expression.

Now look at Figure 3.7, showing the light clock as viewed from the frame it is moving through, and again ask how far apart are the same two events, as measured in this frame. This time, the separation between the events includes both a temporal separation $\Delta t$, and a spatial separation $\Delta x = v\Delta t$. Taking further pythagorean inspiration, let's *suppose* that there is a quasi-pythagorean expression which represents the distance here, and write

$$s^2 = \Delta t^2 + a\Delta x^2,$$

for some constant $a$ we don't yet know (we are at this point taking a very *non*-axiomatic approach to thinking about this, in contrast to the approach we took in Chapter 2). We rewrite Eq. (3.2) using our new convention of $c = 1$, and find

$$\Delta t^2 = (2L)^2 + (v\Delta t)^2 = \Delta t'^2 + \Delta x^2.$$

Combining these, we conclude that

$$s^2 = \Delta t'^2 + \Delta x^2 + a\Delta x^2 = s'^2 + (1 + a)\Delta x^2.$$

Looking at this, we can see that we will have $s^2 = s'^2$ – that is, it will be an *invariant of the transformation* – if $a = -1$, or

$$s^2 = \Delta t^2 - \Delta x^2. \tag{4.3}$$

This is our 'distance function', which has the same property of frame invariance in Minkowski space, that $d^2 = \Delta x^2 + \Delta y^2$ has in euclidean space. It is a difference of squares rather than a sum of squares, and this is the first clue that the *geometry* of Minkowski space – that is, the rules of distance there – is systematically different from euclidean space. The argument in this paragraph, leading up to Eq. (4.3), is founded on the requirement, in Section 3.3, as re-examined here, that the speed of light is measured to be the same in both frames.

This quantity $s^2$ is a frame-invariant separation between two events, where $\Delta t$ and $\Delta x$ are the differences in coordinates of the two events. Since the orientation of our axes is arbitrary, we can immediately generalise Eq. (4.3) into a $(3 + 1)$-dimensional version, where

$$s^2 = \Delta t^2 - (\Delta x^2 + \Delta y^2 + \Delta z^2). \tag{4.4}$$

This quantity is referred to as the *interval*, or sometimes, interchangeably, as the *squared interval* or the *invariant interval*. It is also sometimes written as $\Delta s^2$, but since the interval is always a difference, the $\Delta$ is somewhat redundant. It is also the first appearance of the metric of General Relativity, which we will learn a little more about in Appendix A.

I have chosen to introduce the interval, here, by observing that this particular combination of space and time intervals is frame-invariant. This follows from the axioms we have chosen, and their consequences in terms of time dilation and length contraction. An alternative approach is to simply assert this invariance as an axiom, in place of the constancy of $c$, and deduce the rest of SR from that, in the same way that we could imagine asserting Pythagoras's theorem, and deducing euclidean geometry from there.

Although the interval is what we will mean by 'distance' in Minkowski space, the negative sign in Eq. (4.3) means that the 'distance' has some rather unexpected properties.

Firstly, consider two events which are separated by a time $\Delta t$, but which happen at the same location (so that $\Delta x = 0$; the light clock is an example of this). According to Eq. (4.3) the 'distance' between these events (I'll drop the scare-quotes from now on) is simply $s^2 = \Delta t^2$, and as you can see, this distance will always be positive.

Now consider a flash of light emitted at one event, and received at another a distance $\Delta x$ away. Since the light moves at speed $c = 1$, the second event will take place a time $\Delta t = \Delta x$ later than the first. This means, according to Eq. (4.3), that the distance between the two events is $s^2 = 0$. In other words, *any* two events which are separated by a light flash are, in this geometry, zero distance apart.

Finally, consider two events which are further separated in space than they are in time, so that $\Delta x^2$ is greater than $\Delta t^2$. In this case, the quantity $s^2$ will be negative. It's rather unexpected to find that a distance function can be negative, and we will look at some of the consequences of this below. Note that it is the quantity written '$s^2$' which is the distance function: there is no straightforward interpretation of the quantity '$s$'; instead, we might find ourselves reasoning that, if two events are separated by a distance $s^2 < 0$, then we can conclude that there is a frame $S'$ in which they are separated by $\Delta t' = 0$, in which $|\Delta x'^2| = \sqrt{-s^2}$.

Let's make this a little more concrete, and link it to what we learned in Chapter 3. Specifically, is this consistent with Eq. (3.3)? Let event ① be the light being emitted from the flashbulb at the bottom of the light clock, and event ② the light being received there again, after its round trip. As before, we take the primed frame, $S'$, to be fixed to the clock, and we can decide to put its space origin at the location of the bulb and its time origin at the instant the bulb flashes. This means that the coordinates of the two events are, respectively, $(t_1', x_1') = (0, 0)$ (by the definition of our frame's origin) and $(t_2', x_2') = (2L, 0)$ (this is a restatement of Eq. (3.1) with $c = 1$, combined with the statement that ② happens at the spatial origin). The interval between these two events is therefore

$$s_{12}'^2 = \Delta t'^2 - \Delta x'^2$$
$$= \left(t_2' - t_1'\right)^2 - \left(x_2' - x_1'\right)^2$$
$$= (2L)^2. \tag{4.5}$$

What are the coordinates of these *same* events in the frame $S$, through which frame $S'$ is moving? Because we assume that the two frames are in standard configuration (Section 1.7), we know immediately that an event which happened at the (spatial and temporal) origin of frame $S'$ happened at the origin of frame $S$ also, so the coordinates in frame $S$ of event ① are $(t_1, x_1) = (0, 0)$. For definiteness, let us suppose that the light clock has size $L = 1$ m (and so $s_{12}'^2 = 4$ m$^2$) and is moving at speed $v = 3/5$, so that $\gamma = 5/4$. Thus the interval between the two events is, according to Eq. (4.5), $s_{12}'^2 = 4$ m$^2$ (we shouldn't attempt to interpret this as an 'area', of course, but we will be well-behaved and include the units in numerical expressions below). Equation (3.3) tells us that $t_2 = 2.5$ m (since $t_2' = 2$ m and $\gamma = 5/4$), and Figure 3.7 reminds us that $x_2 = vt_2$, or $x_2 = 1.5$ m. Putting these together, we find

$$
\begin{aligned}
s_{12}^2 &= \Delta t^2 - \Delta x^2 \\
&= (t_2 - t_1)^2 - (x_2 - x_1)^2 \\
&= (2.5\,\text{m})^2 - (1.5\,\text{m})^2 \\
&= 4\,\text{m}^2,
\end{aligned}
$$

equal to $s_{12}'^2$. Thus we see explicitly that the interval $s_{12}^2$ has the *same* value when worked out using the $S$ frame's coordinates, as it has when worked out using the $S'$ frame's coordinates.

More generally, we can see that, in this scenario, $\Delta t = \gamma \Delta t' = \gamma 2L$ and $\Delta x = v \Delta t$ (compare Section 3.3). Thus

$$
\begin{aligned}
\Delta t'^2 - \Delta x'^2 &= (2L)^2 \\
\Delta t^2 - \Delta x^2 &= \gamma^2 (2L)^2 - v^2 \gamma^2 (2L)^2 \\
&= \gamma^2 (1 - v^2)(2L)^2 \\
&= (2L)^2.
\end{aligned}
$$

**Note:** In the discussion in this section, we could with equal justice have started off by defining $s'^2 = -\Delta t'^2$, and we would have ended up with an interval $s^2 = \Delta x^2 - \Delta t^2$ or, more generally, $s^2 = -\Delta t^2 + \Delta x^2 + \Delta y^2 + \Delta z^2$. Choosing one or the other is purely a matter of convention. I slightly prefer the convention $s^2 = \Delta t^2 - \Delta x^2$, because it means that the interval is the same as the 'proper time', which we come to later; but there's very little in it. Arbitrary though it may be, the convention determines the signs of a number of important equations in the text, and *different texts choose different conventions here*, so you should be aware of this if you read around the subject. Most significantly, the definitions of 'timelike' and 'spacelike' in Section 4.7 swap signs. The sum of the signs of the terms is known as the

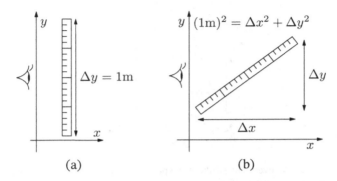

**Figure 4.11** A metre stick in front of you: (a) perpendicular to the line of sight; and (b) at an angle.

*signature*, so that Eq. (4.4), with signs $(+, -, -, -)$, has signature $-2$, and the above alternative definition, with signs $(-, +, +, +)$, has signature $+2$.

[Exercises 4.5 & 4.6]

## 4.5 Changes of Frame, and Perspective

In Figure 4.10 I hoped to illustrate both the arbitrariness of the coordinates of a point, or an event, in two different reference frames, and the commonalities between a rotation and a change of frame in a Minkowski diagram. The closeness of the analogy between these two things is, for me, one of the key insights of this part of our path through SR.

Imagine holding a metre stick in front of you, directly perpendicular to your line of sight (Figure 4.11a). The two ends are, obviously, a metre apart from each other in the $y$-direction. Now partly rotate the stick away from you (Figure 4.11b): the ends of the stick are now *closer together* in the $y$-direction, but separated in the $x$-direction by exactly the right amount to make the pythagorean squared-distance between the two ends of the metre stick, $\Delta x^2 + \Delta y^2 = 1\,\text{m}^2$, as before. You are not in the least bit surprised at this, because you have an intuitive understanding of the invariants of euclidean geometry (although I agree that may not be how you phrase this to yourself at this point). We know that $\Delta x$ and $\Delta y$ here are not frame-independent properties of the measuring rod, but instead a joint property of the rod and our choice of coordinate system.

We can see a very similar thing happening when we look at the separation between two events in Minkowski space, in two frames, as shown in

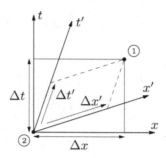

**Figure 4.12** Two events in Minkowski space, and their separations in two reference frames. Compare Figure 4.4.

Figure 4.12. In one frame, the events are separated by $\Delta x$ in space and $\Delta t$ in time, and thus by an interval of $s_{12}^2 = \Delta t^2 - \Delta x^2$. Viewed from a different frame, however, these two events are separated by *different* amounts of space and time, purely because of the change of frame, but the interval between them, $s_{12}^2 = \Delta t'^2 - \Delta x'^2$, is the same.

Exactly as in Figure 4.11(b), the separations along any one coordinate axis are frame-dependent and physically meaningless. It is only the interval $\Delta x^2 + \Delta y^2$, or $\Delta t^2 - \Delta x^2$, that is significant, depending on whether we are talking about the geometry of, respectively, euclidean space (our usual intuition) or Minkowski space (the intuition we must develop for SR).

The correspondence between Figures 4.10(a) and 4.10(b) is not intended to be some vague handwaving analogy. On the contrary, it is a very precise analogy: the same thing is happening in both cases. The only thing that is different between the two cases is the *geometry* of the space in question. In the first case, the transformation from one set of coordinates to the other (Eq. (4.1)) is such that it preserves euclidean distance (Eq. (4.2)). In the second, the coordinate transformation is the one which preserves 'Minkowski distance', Eq. (4.3), with a specific transformation – the expression corresponding to Eq. (4.1) – which we are going to discover in the next chapter. If you think of moving from one inertial frame to another as a 'rotation' (keep the scare-quotes), you will not go far wrong. If you can find a place in your head, where Figure 4.12 looks as natural as Figure 4.11(b), you will have reached some relativistic nirvana.[5]

---

[5] I haven't got there yet.

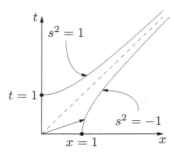

**Figure 4.13** All the points on the two hyperbolae are the same Minkowski-distance from the origin.

## 4.6 Length Contraction and Time Dilation in the Minkowski Diagram

At the cost of some admittedly rather forbidding-looking Minkowski diagrams, it is possible to illustrate time dilation and length contraction graphically. You might want to come back to this section a little later, once you have become more comfortable with Minkowski diagrams, in the next chapter.

In Figure 4.9, we could draw an arc indicating the locus of points $(x, y)$ with the same value of the invariant $r$ (this locus is known, outside of a maths class, as that exotic object the 'Circle'). Can we do a similar thing with the invariant $s^2$ on the Minkowski diagram? Yes: in Figure 4.13 we have drawn the locus of points for which $t^2 - x^2 = +1$ (upper line) and $-1$ (lower). This makes very clear how different is the geometry of a Minkowski diagram from that on the plane in Figure 4.9; all the points on each hyperbola in the figure are the *same* distance $s^2$ from the origin, either $s^2 = 1$ (upper line) or $s^2 = -1$ (lower line).

Imagine a set of clocks which tick every 10 m of time (i.e., every 33 ns in physical units). We put one on the station platform (at $x = 0$) and another in a moving train carriage (at $x' = 0$), and set things up so that there is a tick at $t = t' = 0$. The platform clock will tick again at $t = 10$ m, and the train clock at $t' = 10$ m. The interval between ticks of a particular clock will always be $s^2 = (10\,\text{m})^2 - 0^2$. This is a quite general property: if an inertially moving clock is present at two events (in this case, successive ticks of the clock), then the interval between them is governed entirely by the separation in time in the clock's frame, since in this frame the two events are separated by $\Delta x = 0$; that time interval is what is showing on the clock's face (this idea that the clock face shows the 'distance through spacetime' is what motivates the image of the taffrail log in Section 1.6). This time interval – which is

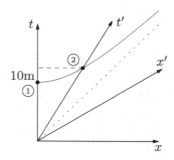

**Figure 4.14** Two clock-ticks on clocks at rest in frames $S$ and $S'$. In each case, the previous tick was at time $t = t' = 0$. The null line is shown dotted.

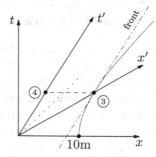

**Figure 4.15** A 10 m long train (at rest in $S'$) moving through a station (at rest in $S$) with a 10 m long platform, also showing (dot-dashed) the worldline of the front of the train.

very immediately related to the Minkowski interval – is known as the *proper time* (we have a little more to say about this in Section 5.4).

Now look at Figure 4.14. The second tick of the platform clock is at time 10 m, at the point marked ①, and the second tick of the train clock is at ② (how do we know that? Because that event is on the $t'$-axis, and the hyperbola marks out the locus of points which are an interval of $s^2 = (10\,\text{m})^2$ from the origin). But the $t$-coordinate of this event ② is *larger* than 10 m – this is time dilation.

Now measure 10 m along the station platform, and mark this point in Figure 4.15. Have a 10 m long train go through the station, timetabled so that the back of the moving carriage is at $x = x' = 0$ at time $t = t' = 0$ (standard configuration again). The worldline of the back of the carriage lies along the $t'$-axis, and the worldline of the front is parallel to this.

Suppose that an observer at the front of the carriage sets off a firework

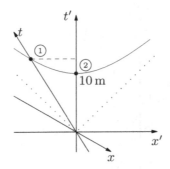

**Figure 4.16** The same events as shown in Figure 4.14, but this time shown in the frame $S'$.

at time $t' = 0$ (that is, simultaneously, in $S'$, with the back of the carriage being at $x = x' = 0$). That front firework will be positioned at event ③. The hyperbola again marks out the set of points which are the same distance – 10 m in this case – from the origin, and simultaneity in $S'$ implies that the event ③ is on the $x'$-axis. How long does the platform party measure this train carriage to be? The way they measure that is to identify where the front of the carriage was at some instant (choosing for example event ③, at the position and time of the firework explosion), and find another event at the back of the carriage *at the same time in S*, namely event ④. Directly from the diagram, you can see that the spatial distance, in $S$, between events ③ and ④, is less than 10 m – this is length contraction.

Now let's look at those events in a Minkowski diagram drawn in the *train's* frame; see Figure 4.16. As before, the worldline of the zero marker on the platform scale lies along the $t$-axis. Events ① and ② are, as before, at positions $(t, x) = (10\,\text{m}, 0)$ and $(t', x') = (10\,\text{m}, 0)$, respectively. The 'calibration' hyperbola again shows that the two events are the same interval away from the origin. In this frame, and unlike Figure 4.14, event ① is later than event ② (in the sense that $t'_1 > t'_2$), so it is the platform's clock which is time-dilated in the train's frame.

If we do the same thing with events ③ and ④, we get Figure 4.17. As before, event ③ is at the front of the train, $x' = 10\,\text{m}$, and the point which the platform observers measure as simultaneous with it, ④, is also shown for reference (just as in Figure 4.15, the line joining ④ and ⑤ is parallel to the $x$-axis). I have also shown a new event ⑤, which is at the 10 m on the platform (as we can see from the calibration hyperbola), and an event ⑥, which is at the zero-end of the station platform (it's on the worldline of the end of the platform, overlaid on the $t$-axis), which happens simultaneously

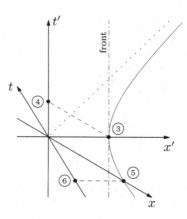

**Figure 4.17** The same events as Figure 4.15, in $S'$.

with ⑤ in the train frame (i.e., $t'_5 = t'_6$); it is clear that the separation between events ⑤ and ⑥ – that is, between the zero and 10 m marks on the platform – is less than 10 m: the platform is length contracted in the train's frame.

Just as we saw at the end of Section 3.2, both length contraction and time dilation are *symmetrical*. The train is contracted in the platform's frame, and the platform in the train's frame; and the train's clocks are dilated or slow in the platform's frame, and the platform's clocks in the train's, all by the same factor, $\gamma$.

We are now ready to derive the Lorentz transformation, but before we do that, there are a couple of other remarks we can make about Minkowski space.                                                              [Exercise 4.7]

## 4.7 Worldlines and Causality

In Figure 4.4, back on p. 53, we saw how we ascribe $(t, x)$ and $(t', x')$ coordinates to an event. In this figure, the $t'$-axis indicates the locus of positions of a particle at the origin of the $S'$ frame, as that frame moves through $S$. This line, which is straight in this case but would be curved for a more general motion of a point, is the worldline of a particle at rest at $x' = 0$. As suggested above, the worldline of a photon, or anything else moving at the speed of light, always appears as a 45° diagonal line on a Minkowski diagram.

Since the interval is just $s^2 = \Delta t^2 - \Delta x^2$, we can see that all the points on a photon's worldline have $s^2 = 0$, all the points below that diagonal have $s^2 < 0$, and all those above it have $s^2 > 0$. Such separations are termed,

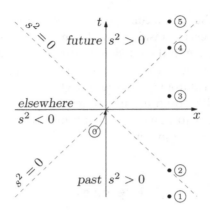

**Figure 4.18** Light cones.

respectively, *null* (or *lightlike*), *spacelike* and *timelike* separations.

We can illustrate these terms in Figure 4.18. Events ⑤ and ① have time-like separations from an event ⓪, at the origin, events ④ and ② have null separations from ⓪, and event ③ has a spacelike separation from ⓪. We can immediately categorise the five events in terms of the type of separation between the event and an event at the origin.

**Event** ⑤ timelike separation from ⓪: a slower-than-light signal could travel from the origin to ⑤.

**Event** ④ null separation: only a speed-of-light signal could make it from ⓪ to ④.

**Event** ③ spacelike separation: no signal could travel between ⓪ and ③, but there is a frame in which ⓪ and ③ are simultaneous.

**Event** ② null separation: a speed-of-light signal could travel from ② to ⓪.

**Event** ① timelike separation: a slower-than-light signal could get from ① to ⓪.

Although we are familiar with time being simply divided into a future and a past, we can see from this Minkowski diagram that spacetime is divided into *three* regions, the familiar 'future' and 'past', plus a region indicated by 'elsewhere', consisting of events which cannot interact causally with events at the origin.

Figure 4.18 shows only the $x$- and $t$-dimensions. If we showed the $y$-dimension as well, and rotated the figure around the $t$-axis, the dashed lines would turn into a *cone* defined by $s^2 = t^2 - (x^2 + y^2) = 0$. This *null cone* demarcates areas of spacetime in the same way as the diagonals in the figure. Every event has a null cone, pointing both forwards and backwards along the time axis: events 'inside' the null cone are timelike separated from it, with $s^2 > 0$, and are in the possible future of the event or its possible past, depending on whether $t > 0$ or $t < 0$; events 'outside' the null cone are spacelike separated from it, and are 'elsewhere'; events on the null cone have a null separation from it, and are reachable only by a light signal. Although it is harder to visualise, the full null cone is the surface $s^2 = t^2 - (x^2 + y^2 + z^2) = 0$ in spacetime.                    [Exercises 4.8 & 4.9]

## Exercises

### Exercise 4.1 (§4.1)

Convert the following to units in which $c = 1$:

1. $1\,\mathrm{J}$
2. $1\,\mathrm{ns}$
3. light-bulb power, $100\,\mathrm{W}$
4. Planck's constant, $h = 6.626 \times 10^{-34}\,\mathrm{J\,s}$
5. jogging pace, $3\,\mathrm{m\,s^{-1}}$
6. momentum of a jogger, $300\,\mathrm{kg\,m\,s^{-1}}$
7. power station output, $1\,\mathrm{GW}$

Convert the following to physical units:

1. velocity $10^{-2}$
2. time $9.46 \times 10^{15}\,\mathrm{m}$
3. acceleration $1.11 \times 10^{-16}\,\mathrm{m^{-1}}$
4. $t' = \gamma(t - vx)$ (this is Eq. (5.6), from the next chapter)
5. the 'mass-shell' equation $E^2 = p^2 + m^2$

If you're having trouble working out whether to multiply or divide, I find it helpful to remember that, just as an inch is a small number of kilometres, a light-metre is a small number of seconds, or a light-second is a large number of metres.

### Exercise 4.2 (§4.1)

What is $9\,\mathrm{N} = 9\,\mathrm{kg\,m\,s^{-2}}$ in natural units?

1. $10^{-16}\,\mathrm{kg}$
2. $10^{-16}\,\mathrm{kg\,m^{-1}}$

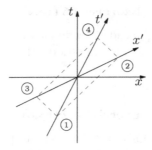

**Figure 4.19** Reflections in the light clock, in the stationary frame.

3. $10^{-16}$ kg m s$^{-2}$

You can also do this without doing any arithmetic – think of the dimensions. [$d^-$]

## Exercise 4.3 (§4.1)

From Exercise 4.2, just above, 9 N is $10^{-16}$ kg m$^{-1}$ in natural units. Given that, work out the acceleration of a mass of 0.1 kg in natural units (you should end up with something in units of m$^{-1}$). What does it mean, in a point of view where speeds are fractions of the speed of light, to have an acceleration in units of per-metre?

## Exercise 4.4 (§4.2)

Which of the following is true *in frame S* in Figure 4.19?

1. Event ③ happens before event ②
2. Event ③ happens at the same time as ②
3. Event ③ happens after ②

And in frame $S'$? [$d^-$]

## Exercise 4.5 (§4.4)

The discussion around Eq. (4.5) showed that $\Delta t^2 - \Delta x^2$ was conserved in that case. Demonstrate that the quantity $\Delta t^2 + \Delta x^2$ is *not* conserved; that is that $\Delta t^2 + \Delta x^2 \neq \Delta t'^2 + \Delta x'^2$.

## Exercise 4.6 (§4.4)

Re-do the calculation at the end of Section 4.4 with a few different speeds, such as $v = 4/5$, or $v = 12/13$, and observe that you get the same value for $s^2$ each time.                                                                     $[d^-u^+]$

## Exercise 4.7 (§4.6)

Cosmic rays entering the top of the Earth's atmosphere collide with it, and produce showers of elementary particles. Of these showers, almost the only 'exotic' particles to reach the ground are muons. These muons have a half-life of $t_{1/2} = 2.2\,\mu s$, and are travelling at relativistic speeds with, on average, a $\gamma$ of 40. The majority of the produced muons reach the ground, where they can be detected.

(a) Supposing that the muons are created at a nominal height of 15 000 m, calculate how long it takes them to reach the ground, in the Earth frame, giving the answer in seconds and in half-lives.

(b) Calculate how long it takes the muons to reach Earth, in the muons' frame, giving the answer, again, in seconds and half-lives.

(c) Can this be regarded as a test of Special Relativity?

(d) In the frame of the muons, the Earth is moving towards them at (approximately) the speed of light. Calculate the distance between the altitude at which the muons are created, and the surface of the Earth, in the muons' frame, and thus how long it takes for the muons to traverse it, giving your answer, again, in seconds and half-lives.

(e) Comment on the equality or inequality of your answers to parts (b) and (d).

## Exercise 4.8 (§4.7)

The usual train travels through the usual station at a speed $v = 1/5$, in units where $c = 1$. The train guard sets off a firecracker 500 m in time after the train has passed the station, when the train is 100 m further down the track (both distances measured in the station's frame). Draw a Minkowski diagram, indicating on it the worldlines of the station and the train, and the explosion event. At what time is the explosion visible at the station?

## Exercise 4.9 (§4.7)

Consider three events, with coordinates $(t_1, x_1) = (1, 1)$, $(t_2, x_2) = (6, 4)$ and

$(t_3, x_3) = (4, 6)$ (take $c = 1$). Calculate the intervals $s_{12}^2$, $s_{13}^2$ and $s_{23}^2$ between the three pairs of events, state whether each is timelike, spacelike or null, and in each case whether the earlier event could influence the later event through a signal travelling no faster than light.

# 5

# The Lorentz Transformation

In Section 3.2 we saw how observers could make measurements of lengths and times in frames which are in relative motion, and reasonably disagree about the results – the phenomena of length contraction and time dilation. In Section 3.3, we were able to put numbers to this and derive a quantitative relation, Eq. (3.3), between the duration of a 'tick' of the light clock as measured in two frames. We want to do better than this, and find a way to relate the coordinates of any event, as measured in any pair of frames in relative motion. That relation – a transformation from one coordinate system to another – is the *Lorentz transformation* (LT). The derivation in Section 5.1 has a lot in common with the account given in Rindler (2006).

Most of the work of this chapter is in Section 5.1. The rest is, to a greater or lesser extent, commentary on that, with several sections being marked with dangerous bends, and thus omissible.

*Aims*: you should:

5.1. understand the derivation of the Lorentz transformation, and recognise its significance.

## 5.1 The Derivation of the Lorentz Transformation

Consider two frames in standard configuration, and imagine an event such as a flashbulb going off; observers in each of the two frames will be able to measure the coordinates of this event. Those observers will of course produce different numbers for those coordinates – they will disagree about the precise time and location of the event – with those in frame $S$ producing coordinates $(t, x, y, z)$, and those in $S'$ producing $(t', x', y', z')$. It is our task now to calculate the relationship between those two sets of numbers.

First of all, we can note that $y' = y$ and $z' = z$, since this is just a re-statement of the lack of a perpendicular length contraction, as discussed in Section 3.3. Therefore we can unproblematically make things easier for ourselves by supposing that the event takes place on the x-axis, so that $y = z = 0$ (problem: construct for yourself the argument that there is no loss of generality here). Therefore our problem has reduced to that of obtaining $(t', x')$ for this event, given $(t, x)$ (or vice versa).

Now imagine a second event, located at the origin with coordinates $(0, 0)$ in frame $S$; since the frames are in standard configuration, we can immediately deduce that this event has coordinates $(0, 0)$ in frame $S'$ also. Since we have two events, we have an interval between them, with the value $s^2 = (t - 0)^2 - (x - 0)^2 = t^2 - x^2$ in frame $S$. We will now *assume* that the interval is frame-independent (recall Section 4.4): the calculation of this interval done by the observer in the primed frame will produce the same value, restating Eq. (4.3):

$$t^2 - x^2 = s^2 = t'^2 - x'^2. \tag{5.1}$$

Thus the relationship between $(t, x)$ and $(t', x')$ must be one for which Eq. (5.1) is true.

Equation (5.1) is strongly reminiscent of Eq. (4.2), and we can make it more so by writing $l = it$ and $l' = it'$, so that Eq. (5.1) becomes[1]

$$l^2 + x^2 = -s^2 = l'^2 + x'^2. \tag{5.2}$$

But we already know a transformation which satisfies this constraint, and so can conclude that the pairs $(l, x)$ and $(l', x')$ can be related *via* the analogue of Eq. (4.1), and thus that

$$x' = \quad x \cos \theta + l \sin \theta \tag{5.3a}$$
$$l' = -x \sin \theta + l \cos \theta, \tag{5.3b}$$

for some angle $\theta$, which depends on $v$, the relative speed of frame $S'$ in frame $S$. That is, this specifies a linear relation for which Eq. (5.1) is true. If we finally write $\theta = i\phi$,[2] and recall the trigonometric identities $\sin i\phi =$

---

[1] You may wish to refer to Appendix C, to refresh your memory on complex numbers, and trigonometric functions using them.
[2] If $\theta$ is pure imaginary then, from the Taylor expansions, $\cos \theta$ is real and $\sin \theta$ is pure imaginary, so that $x'$ and $l'$ are real and pure imaginary, respectively, as they should be.

$i \sinh \phi$ and $\cos i\phi = \cosh \phi$, then

$$t' = \quad t \cosh \phi - x \sinh \phi \qquad (5.4a)$$
$$x' = -t \sinh \phi + x \cosh \phi \qquad (5.4b)$$

(I have swapped the order of the two equations, relative to Eq. (5.3), and swapped $x$ and $t$ in each row, for later notational convenience; you might want to take a look at Exercise 6.5 for an alternative way of arriving at these equations.)

Now consider an event at the spatial origin of the moving frame, that is, at $x' = 0$ for some unknown $t'$. What are the coordinates of this event in the unprimed frame? That's easy: if it happens at time $t$ in the unprimed frame then it happens at position $x = vt$ in that frame (because the two frames are in standard configuration), in which case Eq. (5.4b) can be rewritten as

$$\tanh \phi(v) = v. \qquad (5.5)$$

The quantity $\phi(v)$ is sometimes called the 'rapidity', and the application of an LT to a frame is sometimes called a 'boost' of that frame.

Since we now have $\phi$ as a function of $v$, we also have, in Eq. (5.4), the full transformation between the two frames; but combining Eq. (5.4) and Eq. (5.5) with a little hyperbolic trigonometry (remember $\cosh^2 \phi - \sinh^2 \phi = 1$), we can rewrite Eq. (5.4) in the more usual form

$$t' = \gamma(t - vx) \qquad (5.6a)$$
$$x' = \gamma(x - vt). \qquad (5.6b)$$
$$y' = y \qquad (5.6c)$$
$$z' = z, \qquad (5.6d)$$

where we have included the trivial transformations for $y$ and $z$, which we deduced above. Here $\gamma$ is (as in Eq. (3.4) but now with $c = 1$; see also Section C.2)

$$\gamma(v) = \cosh \phi = \left(1 - v^2\right)^{-1/2}. \qquad (5.7)$$

If frame $S'$ is moving with speed $v$ relative to $S$, then $S$ must have a speed $-v$ relative to $S'$. Swapping the roles of the primed and unprimed frames, the transformation from frame $S'$ to frame $S$ is exactly the same as Eq. (5.6), but with the opposite sign for $v$:

$$t = \gamma(t' + vx') \qquad (5.8a)$$
$$x = \gamma(x' + vt'), \qquad (5.8b)$$

which can be verified by direct solution of Eq. (5.6) for the unprimed coordinates. [Exercises 5.1 & 5.2]

## 5.2 Addition of Velocities

Adding and subtracting the expressions in Eq. (5.4), and recalling that $e^{\pm\phi} = \cosh\phi \pm \sinh\phi$, we find

$$t' - x' = e^{\phi}(t - x) \tag{5.9a}$$
$$t' + x' = e^{-\phi}(t + x), \tag{5.9b}$$

as yet another form (once we add $y' = y$ and $z' = z$) of the LT.

Consider now three frames, $S$, $S'$ and $S''$. If $S$ and $S'$ are in standard configuration with relative velocity $v_1$, and $S'$ and $S''$ are in standard configuration with relative velocity $v_2$ (implying that the direction of velocity $v_1$ is parallel to that of $v_2$), then frames $S$ and $S''$ will also be in standard configuration with some third velocity $v$; we obviously cannot presume that $v = v_1 + v_2$, as it would be under a galilean transformation. However, applying Eq. (5.9a) twice shows us that

$$t'' - x'' = e^{\phi}(t - x) = e^{\phi_1 + \phi_2}(t - x), \tag{5.10}$$

where $\phi$, $\phi_1$ and $\phi_2$ are the rapidity parameters corresponding to $v$, $v_1$ and $v_2$. This shows us how to add velocities: Eq. (5.5) plus a little more hyperbolic trigonometry ($\tanh(\phi_1 + \phi_2) = (\tanh\phi_1 + \tanh\phi_2)/(1 + \tanh\phi_1 \tanh\phi_2)$) produces

$$v = \frac{v_1 + v_2}{1 + v_1 v_2}. \tag{5.11}$$

We used Eq. (5.10) as a stepping-stone on our way to the velocity-addition formula. However it also demonstrates the important point that, if frames $S$ and $S'$ are related by a LT, and frame $S'$ is related to a further frame $S''$ by another LT, then there is also necessarily a single LT which takes us from $S$ to $S''$. This is another hint that length contraction and time dilation are not simply isolated consequences of a change of speed, but glimpses of a larger underlying structure. [Exercises 5.3 & 5.4]

### 5.2.1 The Lorentz Transformation as a Group ⚠

The form of the LT shown in Eq. (5.9), and the addition law in Eq. (5.10), together conveniently indicate three interesting things about the LT: (i) for any two transformations performed one after the other, there exists a third with the same net effect (i.e., the LT is 'transitive'); (ii) there exists a transformation (with $\phi = 0$) which maps $(t, x)$ to themselves (i.e., there exists an identity transformation); and (iii) for every transformation (with $\phi = \phi_1$, say) there exists another transformation (with $\phi = -\phi_1$) which results in the identity transformation (i.e., there exists an inverse). If we add to these three properties the observation that, for three Lorentz transformations $L_1$, $L_2$ and $L_3$, the associative property holds, $L_3 \circ (L_2 \circ L_1) = (L_3 \circ L_2) \circ L_1$, then these are enough to indicate that the LT is an example of a mathematical 'group', known as the 'Lorentz group'.

The Lorentz group consists of all those transformations which leave invariant the interval in Eq. (4.4) (or, as we shall see in the next chapter, which leave the lengths of 4-vectors invariant). The LT of Eq. (5.6) describes transformations where the motion is along the $x$-axis (i.e., standard configuration). It is possible to generalise this transformation to boosts in any spatial direction, and to include spatial rotations (it's usual to use one of a few different notations when discussing this, to avoid the discussion becoming unbearably messy); this is referred to as the group of *restricted (homogeneous) Lorentz transformations*. The 'restriction' is that this excludes reflections in space or time, and the full Lorentz group consists of the restricted group extended with transformations which flip the signs of the coordinates. If we further add translations – transformations which change the origin of the coordinate system – we obtain the 'inhomogeneous Lorentz transformations', also known as the *Poincaré transformations*, which represent the *Poincaré group*.

Groups are very closely related to the mathematical study of symmetry (to say a sphere is 'symmetric' is to say that there is a mathematically analysable set of rotations you can perform, which leave the sphere looking unchanged). Einstein was repeatedly motivated by ideas of symmetry, and invariance, and they are of profound importance in modern physics.

Poincaré has a lot to do with the mathematical history of relativity. You might also be interested in a discussion (Adlam 2011) of how much Poincaré understood of what we now know as relativity, and how his approach to relativity is linked to a particular philosophical position on science (this is *firmly* in the category of dangerous-bend remarks).

## 5.3 The Invariant Interval and the Geometry of Spacetime

Is the interval merely a mathematical curiosity? No; it is the key to a profound reappraisal of our picture of space and time.

The LT as represented by Eq. (5.4) *looks* like a rotation – indeed like the rotation which Eq. (5.3) would represent were $l$ and $\theta$ real. The hyperbolic functions and the pattern of signs tell us that it is not exactly the same, but the similarities are instructive.

A crucial – indeed the defining – feature of a rotation in a plane is that it preserves euclidean distance. If you have a point $(x, y)$ in the plane, and you rotate the axes so that the same point now has coordinates $(x', y') = (x \cos\theta + y \sin\theta, -x \sin\theta + y \cos\theta)$, then you know that the distance $x^2 + y^2$ will be equal to $x'^2 + y'^2$. There's nothing which marks out one pair $(x, y)$ as more fundamental in any sense than another pair $(x', y')$, nor even any important difference between the $x$- and $y$-axes – they're just a pair of directions on the plane. The only fixed thing here is the euclidean distance.

In the same way, the crucial feature of the LT is that it preserves the interval $s^2$. Just as in the previous paragraph, there is nothing fundamental about the particular pair of coordinates $(t, x)$ which our clocks and measuring rods pick out for us; the coordinates $(t', x')$ which are natural for another, moving, observer, are just as fundamental. Similarly, and surprisingly, there is no real difference between the $t$- and $x$-axes – they're just different directions in the relevant geometrical space. This is really an astonishingly radical view of space and time, and drove Minkowski to remark, famously, that

> The views of space and time which I wish to lay before you have sprung from the soil of experimental physics, and therein lies their strength. They are radical. Henceforth space by itself, and time by itself, are doomed to fade away into mere shadows, and only a kind of union of the two will preserve an independent reality. (Minkowski 1908)

In the newtonian world-view, Space was a three-dimensional geometrical space, which obeyed Euclid's geometrical axioms, and specifically preserved the euclidean length. Time entered in a simple way: history consisted of this three-dimensional space moving gracefully through Time, with each instant of history having its 'own' space associated with it. Special Relativity teaches us that the world is not like this after all: instead we live in a *four*-dimensional universe, with three space dimensions and one time dimension – *Spacetime*. Spacetime does not obey euclidean geometry, since a 'rotation' in spacetime (i.e., a Lorentz transformation) preserves, not euclidean distance, but the interval $s^2$. The geometry of spacetime contains a great deal more structure than this, which you will learn much more about if you study General Relativity.

A useful picture, I find, is to imagine two cars driving roughly northwards in the desert, at the same speed: $A$ is moving north-northwest and $B$ is moving north-northeast; thus $A$ is moving 'forwards' (i.e., NNW) faster than $B$ is moving towards the NNW, and $B$ is moving 'forwards' (i.e., NNE) faster than $A$ is moving towards the NNE. That is, each of the two cars is moving forwards steadily and continuously both faster *and* slower than the other – this apparent paradox evaporates, and is revealed as the result of careless language, as soon as one realises that the two cars have different ideas of 'forward', because of the different orientations of their reference frames – one reference frame is rotated with respect to the other. Exactly analogously, remember that at the end of Section 3.2 we realised that the observers in the two trains each thought that the other was moving through time more slowly than they were: just as with the cars in the desert, this is perfectly true, because the two sets of observers have different notions of 'forwards in the desert' or 'forwards in time' as a result of their systematically 'rotated' reference frames. This picture seems to me to be the most concrete illustration of the relationship between the passage of time, and the geometry of spacetime.

## 5.4 Proper Time and the Invariant Interval

In some frame $S$, consider two events, one at the origin, and one at coordinates $(t, x)$. The frame is arbitrary – there is nothing special about the numbers $x$ and $t$ obtained in this frame.

Consider now a clock moving at a constant velocity $v$ in $S$, chosen so that it is present at both events (we assume for the moment that the two events are close enough that this is possible for a clock moving at a sub-light speed). Of all the time and space coordinates which could be used to describe the motion of this clock, there is one we can pick out as special: the frame, $S'$, in which the clock is at rest at the origin. In this frame, the clock's position is constant, $x' = 0$, because both events happen at the origin, and the time coordinate $t'$ is the time which the clock actually shows on its face. We write this particular time coordinate as $\tau$, and call it the *proper time*. Starting from the frame $S$ in which the clock has coordinates $(t, x)$, we can use Eq. (5.6a) to transform to the clock's rest frame and write

$$\tau = \gamma(t - vx). \tag{5.12}$$

The frame $S'$ is special – it's the one where the clock is at rest – but the frame $S$ is arbitrary, and no matter how frame $S$ is moving with respect to the clock, it would always be possible to calculate this number $\tau$, showing on the face of the clock. That proper time is invariant under a LT.

Noting that the velocity required for this scenario is simply $v = x/t$, we can put this velocity into Eq. (5.12) to find $\tau^2 = t^2 - x^2$. At this point we realise that we have discovered another version of Eq. (4.4), and that the idea that the proper time is an easily graspable proxy for the interval is what is behind much of the discussion in Section 4.6.

## 5.5 Applications of the Lorentz Transformation

The applications of the LT are basically all the same: given a set of coordinates in one frame, what are the coordinates of the same event or events in another, moving, frame? I provide a 'recipe' for doing relativity problems in Section D.2.

### 5.5.1 Length Contraction

A rod of length $L_0$ lies along the $x'$-axis of frame $S'$, with one end at the origin; what is its length as measured in the $S$ frame? Frames $S$ and $S'$ are of course in standard configuration.

At time $t = 0$, there will be two observers in $S$ who are adjacent to the ends of the moving rod: they let off two bangers, or firecrackers; call these events ① and ②. We can find the length of the rod by considering the coordinates of these two events in the two frames (this is the definition of the length of the rod, as measured in the 'stationary' frame – see Section 1.4).

In frame $S'$, there is no complication: *any* events which happen at the ends of the rod will be a distance $L_0$ apart, so that $x'_2 = L_0$.

In $S$, the two events happen at $t$-coordinates $t_1 = t_2 = 0$, and $x$-coordinates $x_1 = 0$ and $x_2 = L$. Since the origins were coincident at $t = t' = 0$, we know that $x_1 = x'_1 = 0$ (standard configuration again). From Eq. (5.8), we can write down that

$$x_2 = \gamma(x'_2 + vt'_2) \tag{5.13a}$$

$$0 = t_2 = \gamma(t'_2 + vx'_2). \tag{5.13b}$$

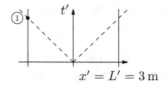

$$x' = L' = 3\,\text{m}$$

**Figure 5.1** Events in the carriage frame, $S'$.

Equation (5.13b) tells us that $t'_2 = -v x'_2$; substituting this into Eq. (5.13a) and writing $x_2 = L$ and $x'_2 = L_0$, we have

$$L = \gamma(1 - v^2)L_0 = \frac{L_0}{\gamma}, \tag{5.14}$$

showing that the moving rod is measured, by observers in the stationary frame, to be shorter than its 'rest length' $L_0$.

Compare Section 3.4. The only difference between this section and that one is that here, we were able to short-circuit the rather elaborate argument of Section 3.4, since the LT in a sense provides the machinery for that argument in a ready-to-use form.                     [Exercise 5.5]

### 5.5.2 A Worked Example: The Trains Again

As in Section 3.1, consider a train moving past a station platform, this time at a specific speed $v = 0.5$ (this is the same setup as in Figure 3.2, but going at a different speed); the carriage is 6 m long, and a flashbulb is set off in the centre of the carriage at time $t' = 0$. (i) What time does the flash reach the back of the carriage as measured in the carriage? (ii) And as measured on the platform? (iii) At the instant when the flash reaches the back of the carriage, what is the position of the forwards-moving wavefront, as measured on the platform? (iv) And as measured in the carriage? (v) Where is the front of the carriage at this instant (measured in the platform's frame)? (vi) So what is the reading, at this instant, on the clock attached to the front of the train? We will work through this using the recipe of Section D.2.

Firstly, we have to identify the frames we want to use, and draw the Minkowski diagram. Let the platform's frame be $S$ and the carriage's be $S'$, with the two frames in standard configuration, and choose the origin of the $S'$ frame to be at the centre of the carriage (the origin of $S$, on the platform, is arbitrary). In drawing the Minkowski diagram(s) you will naturally identify events. The worldlines of the ends of the carriage are drawn in Figure 5.1. Event ① is the flash reaching the back of the carriage, and so sits on both

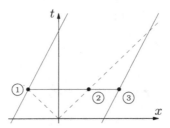

**Figure 5.2** Events in the platform frame, $S$.

that worldline and the worldline of the light flash from the origin, and thus
at the intersection of those two worldlines.

In the platform frame, we draw the worldlines of the ends of the carriage,
in motion to the right, and the worldlines of the same light flashes, again at
slope 1; see Figure 5.2. Exactly as before, event ① is on the intersection of
the same two worldlines.

Question (i): This one's easy: we're looking for the time coordinate of
event ① in the moving frame, thus $t_1'$. We've implicitly been told the position
of this event, and if we take the length of the carriage, in its own frame,
to be $2L' = 6$ m (thus defining $L'$ in a temporarily convenient way), then
$x_1' = -L' = -3$ m. The flash goes off at time $t' = 0$ and moves backwards
and forwards at speed $c = \pm 1$. The back of the carriage is at coordinate
$x_1' = -3$ m, so the flash reaches the back at time $t_1' = -x_1' = 3$ m. At this
point, we've successfully written down most of what the question is telling
us (Section D.2 tells us to 'write down what you know').

(ii) This question translates into a request for $(t_1, x_1)$, given that we now
know $(t_1', x_1')$. The inverse LT is, from Eq. (5.8), $t = \gamma(t' + vx')$, $x = \gamma(x' + vt')$, where

$$\gamma(v = 0.5) = \frac{1}{\sqrt{1 - v^2}} = \frac{1}{\sqrt{1 - 1/4}} = \frac{2}{\sqrt{3}}.$$

Therefore we have $x_1 = -\sqrt{3}$ m and $t_1 = +\sqrt{3}$ m (we can spot that $|x_1| = |t_1|$, indicating that the signal from the origin to event ① is travelling at
$c = 1$, exactly as it should be – that's reassuring).

(iii) In order to associate an event with this part of the question, we
imagine one which happens on the worldline of the light flash, which also
happens at the same time $t$ as event ① (it's hard to imagine a physical event
which happens just here, which is incidentally telling us that this is a rather
artificial question to ask, but the point is that we must synthesise such an
event in our heads, in order to have something to attach coordinates to).

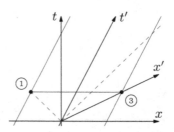

**Figure 5.3** The final diagram, showing both frames.

Call this event ②. Thus we can draw it onto Figure 5.2 on the forwards-going light flash worldline, and simultaneous with event ① (i.e., at the same height as ① on the diagram). The flash started at the origin $x = 0$ of the $S$ frame at time $t = 0$ and moves at speed $c = 1$, so at time $t_2 = t_1 = \sqrt{3}$ m the wavefront must be at position $x_2 = +\sqrt{3}$ m.

(iv) We need the coordinates of event ② in the moving frame; that is, we want $(t'_2, x'_2)$ given that we now have $(t_2, x_2)$. From Eq. (5.6), $x'_2 = \gamma(x - vt) = 1$ m and $t'_2 = \gamma(t - vx) = 1$ m (since the flash is moving at speed $c = 1$, we must have $x'_2 = t'_2$, so this result is reassuring).

(v) Saying 'at this instant (measured in the platform's frame)' tells us that we are looking for a $t$ and not a $t'$. We're looking for the coordinates of an event ③ which is such that $t_3 = t_1$, and the position of which is at the front of the carriage. Marking this in Figure 5.2 confirms to us that all three events have been described as being simultaneous in the platform frame. With $L'$ as defined above, we can reason that the front of the carriage is at (constant) coordinate $x' = L' = 3$ m in the $S'$ frame and so at coordinate $x = L = L'/\gamma$ at time $t = 0$ (by length contraction or from Eq. (5.6b)). Therefore the front of the carriage, moving at speed $v$, is afterwards at coordinate $x = L + vt$. At time $t_3 = t_2 = \sqrt{3}$, the coordinate of the front will therefore be $x_3 = (3\,\text{m})/\gamma + \sqrt{3}/2 = 2\sqrt{3}$ (alternatively, we can spot that we know $x'_3$ and $t_3$, and want $x_3$, so we can use the LT $x'_3 = \gamma(x_3 - vt_3)$ and rearrange for $x_3$; this is more direct, but you will often have to do simple speed-times-time reasoning of this sort).

(vi) Using the result of part (v), we immediately obtain $t'_3 = \gamma(t_3 - vx_3) = 0$, from Eq. (5.6a). Alternatively, we have $t_3 = \sqrt{3}$ m and $x'_3 = 3$ m and want $t'_3$; we can therefore rearrange Eq. (5.8a) to find $t' = t/\gamma - vx'$ and get the same result without going via step (v).

I've drawn both frames in the same diagram, in Figure 5.3. Notice that event ③ is on the $x'$-axis, reassuringly consistent with having coordinate

$t'_3 = 0$. It happened at the same time, in $S'$, as the light flash.

Note that I've drawn Figure 5.2 and Figure 5.3 in their final 'fair-copy' version. Working through this example on paper, I added the various events to the diagram as I progressed through the various questions, and the resulting diagram was a bit of a mess. It's only in retrospect that it becomes clear how to draw the final tidy diagram.

This is quite an intricate calculation (made rather more intricate by my indicating some alternative routes, here and there). Note, however, that we used the LT only in parts (ii), (iv) and (vi). In parts (i), (iii) and (v) we used nothing more sophisticated than 'distance is speed times time', plus a bit of careful thought about what distance and what time we meant. Although this seems fiddly, it ends up being quite routine with practice, as long as you use a systematic approach such as that of Appendix D, and move step by step, from what you know to what you are to calculate.

[Exercises 5.6 & 5.7]

## 5.6 The Equations of Special Relativity in Physical Units

Back in Section 4.1 we introduced 'natural units', where we measure both distance *and time* in units of metres. This makes a lot of SR's equations look both neater and more symmetrical. Here, however, we'll repeat some of the key equations of SR in physical units, partly (a) so that you've seen them at least once, as you might see them in other texts, and (b) to perhaps persuade you that they are unattractively messy in this form.

The interval (see Eq. (4.3)):

$$s^2 = \Delta t^2 - \Delta x^2 \qquad \longrightarrow \qquad s^2 = c^2 \Delta t^2 - \Delta x^2.$$

The Lorentz transformation (see Eq. (5.6)):

$$t' = \gamma(t - vx) \qquad \longrightarrow \qquad ct' = \gamma\left(ct - \frac{vx}{c}\right)$$

$$x' = \gamma(x - vt) \qquad \longrightarrow \qquad x' = \gamma(x - vt)$$

$$\gamma = \left(1 - v^2\right)^{-1/2} \qquad \longrightarrow \qquad \gamma = \left(1 - \frac{v^2}{c^2}\right)^{-1/2}.$$

And the addition of velocities (see Eq. (5.11)):

$$v = \frac{v_1 + v_2}{1 + v_1 v_2} \qquad \longrightarrow \qquad v = \frac{v_1 + v_2}{1 + v_1 v_2 / c^2}.$$

Each of these expressions was obtained from the corresponding expression in natural units by the simple method of adding enough powers of $c$ in each term, that the expression ends up dimensionally correct. Another way is to replace $t \longmapsto ct$ and $v \longmapsto v/c$.

## 5.7 Paradoxes

The teaching of SR conventionally and usefully includes various paradoxes. In this context, 'paradox' means something that seems wrong at first sight, but isn't. These so-called paradoxes are thought experiments where we arrive at conclusions which are probably unexpected, but which are simply correct deductions from ideas we at least partly understand. We look at them in order to deepen our understanding of, and familiarity with, the ideas of SR.

### 5.7.1 The Twins

This is a famous one.

Firstly, imagine Odysseus sailing (at a relaxed non-relativistic speed) between two points, say from Troy to Ithaca. The direct route would follow the $y$-axis in Figure 5.4(a). Contrast this with a route which takes a diversion via Circe, producing the dog-leg route in the figure. Taking this to be a flat euclidean surface, a straight line is the shortest distance between two points, so that the direct route to Ithaca is, of course, shorter than the indirect one. If, in order to measure the distance sailed beyond any doubt, Odysseus towed a taffrail log behind him through the wine-dark sea, it would turn a smaller number of times if he took the direct route, than if he took the indirect one. So far, so obvious.

Now imagine two intrepid scientists, Odysseus and Penelope, who are conveniently twins. Penelope leaves Earth in a spaceship to travel to a star 25 light-years away, and does so at a speed $v = 0.5$; then turns round and comes back. How long is she away for as timed by Odysseus on Earth, and how long in her own frame? If she travels a total distance of 50 light-years at a speed $v = 0.5$, then the journey will take 100 years as measured by Odysseus. However, we know that the time measured on board the spaceship will be

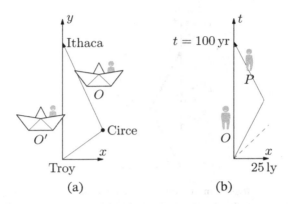

**Figure 5.4** Travellers' paths: (a) Odysseus can go from Troy to Ithaca by two different routes; (b) Penelope travels outward at speed $v = 0.5$ for 25 yr, and then returns.

less, according to Eq. (3.3) above: for $v = 0.5$, we have $\gamma = 1.15$, and only 87 years will pass for Penelope, who will consequently be substantially younger than Odysseus when she returns to Earth.

That seems a little odd, but we're used to peculiarity in SR, now. However, some bright spark then points out that, relative to Penelope, it is *Odysseus* who has been moving, so shouldn't the whole situation be symmetrical, just as with the trains in Section 3.1 above, forcing us to conclude that Odysseus will be younger than Penelope? This is nonsensical – although two trains can be mutually measured each to be shorter than the other, two clocks (Penelope and Odysseus) can't logically each be showing less time than the other.[3]

The paradox is dissolved as soon as we point out that Penelope *cannot* conclude that she is not moving (and thus that Odysseus is moving), since of the two only Penelope experiences the *change* of velocity at the distant star, with an absolutely observable acceleration – SR discusses the relationships between inertial frames, and Penelope does not (indeed cannot) remain in a single inertial frame for the entire round trip. This becomes clearer when we consider the Minkowski diagram.

In Figure 5.4(b), we see the path which Penelope takes through spacetime, on the way to the star and back. It appears that Penelope's route is the longer,

---

[3] The two situations are not as directly comparable as may at first appear. A measurement of distance necessarily involves events separated in space, so we must deal with questions of simultaneity. In contrast, observations of clocks represent a sequence of observations of a single object as it moves through time and space.

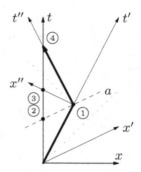

**Figure 5.5** The dog-leg in spacetime, annotated, showing Odysseus's rest frame $S$, and the rest frame of Penelope on the outward ($S'$) and return ($S''$) journeys. The origin of the frame $S''$ is at event ①, and the dashed line $a$ is parallel to the $x'$-axis. The dotted lines are the paths of light flashes moving in the positive and negative $x$ direction.

but remember that the plane of the Minkowski diagram is *not* a euclidean surface, so that our intuitions about lengths and angles are not reliable. It is possible to show, in fact, that a straight line is the *longest* distance between two points in Minkowski space (see Exercise 5.8), so that Odysseus, taking the straight route, has travelled a greater distance through spacetime than has Penelope. Since, for a pair of timelike-separated events, the distance through spacetime, $\sqrt{s^2}$, is the same as the proper time $\tau$ between the events (Section 5.4), we can say that the proper time along Odysseus's path is greater than the proper time along Penelope's, which is to say that a clock carried with him will show a greater time elapsed than one carried by Penelope, which is to say that Odysseus is older than Penelope when they re-meet.

It is at this point that I, for one, get the most immediate benefit from the analogy of the clock as a spacetime version of the taffrail log. When Odysseus sailed, in a boat, from Troy to Ithaca via Circe, the taffrail log showed the amount of Mediterranean he moved through; it would have turned fewer times if he'd gone home by the direct route. When Odysseus and Penelope move from take-off to event ④, their clocks show how much spacetime they have moved through. Clock-Penelope moved through less; she is younger.

Another way of approaching the puzzle is to ask what events are simultaneous in the various different frames. In Figure 5.5 we see the same path through spacetime as in Figure 5.4(b), but this time including the axes of the $S'$ frame (in which Penelope is at rest, on the outward journey; she is thus moving along the $t'$-axis in that frame), and the $S''$ frame (in which she is at rest on the return journey, moving along the $t''$-axis; note that frame $S''$ is unusually *not* in standard configuration with respect to the other two frames). This diagram makes it very clear that Penelope *changes inertial frame* at the turnaround event, ①. We also see that, just before the turnaround, Penelope sees[4] event ②, which is at Odysseus's location, $x = 0$, as being simultaneous with ① in $S'$ (the line $a$ is parallel to the $x'$-axis). Immediately *after* the turnaround, however, she sees event ③ as simultaneous with event ①; in this frame, event ② happened in the past, with respect to ①. If Penelope has skipped her relativity lectures, and is not aware of the significance of the change of frame, she will simply 'miss' the time interval between events ② and ③. During the two 'cruise' phases, before and after the turnaround, the two observers' clocks are indeed symmetrically slower than each other, but the frame-independent difference between the two observers' elapsed time is attributable to the gap between ② and ③. This gap, one might say, is where Odysseus's extra ageing comes from.

There is an alluring blind alley here, prompted by the presence of the acceleration at the turning-point, even prompting some folk to insist that the 'Twins Paradox' needs GR to resolve it. It does not: there is no need to talk about any actual acceleration: rather than actually turning round at the star, Penelope could simply set the clock of another traveller she meets there, who is already travelling at the right speed in the return direction, or hand over her log-book to them. Our conclusion would then be to do with the total elapsed time on the two legs of the journey, as calculated from the log-books, rather than counting the grey hairs on one miraculously unaged traveller; but this is exactly the same conclusion as above, merely in a less vivid form.

As a final remark, a real Penelope would of course accelerate towards the destination, and then slow to a halt before accelerating in the opposite direction for the return journey. Gourgoulhon (2013, §2.6, where he refers to the scenario by its alternative name of *Langevin's traveller*) works through this more realistic case in detail. This doesn't change the logic of the argument, but firstly it illustrates the way that SR can comfortably deal with acceleration (we touch on this also in Exercise 6.11); and secondly it may reassure you that, when we introduce the simplification of an instantaneous turnaround, above, we are not also introducing a subtle flaw in the argument. [Exercise 5.8]

---

[4] More precisely, and to avoid the word 'sees', Penelope can arrange a network of friends at rest in frame $S'$, one of whom ends up local to event ②, and logs it as happening at time $t'_2$, the time of event ② in frame $S'$, which is numerically equal to $t'_1$.

## 5.7.2 The Pole in the Barn

A farmer with a 20 m pole holds it horizontally and runs towards a barn which is 10 m deep.[5] The farmer's wife, standing by the barn door, sees him running at a speed (specifically $v = \sqrt{3}/2$) at which $\gamma = 2$. The pole is therefore length contracted to have a measured length of 10 m in the barn's frame, so that the pole will fit entirely into the barn, and the farmer's wife can slam the door behind him, with the '20 m' pole entirely (and briefly!) within the 10 m barn (in other words, the farmer's wife is 'measuring the length of the pole' using yet another variant of the procedure described in Section 1.4).

However, in the farmer's frame, the barn is rushing towards him with the same $\gamma$-factor, so that the barn is measured to have a length of only 5 m, making it even more outlandish that the pole should fit in the barn.

The reason this sounds so bizarre is that we are forgetting the central insight of this chapter: spatially separate events which are simultaneous in one frame will not be simultaneous in another. When we say 'the pole is entirely within the barn' we *mean* something like 'there is at least one pair of events, located at each end of the pole while those ends are within the barn, which are simultaneous in the frame of the barn'. However, this pair of events are *not* simultaneous in the frame of the pole. In other words, the paradox comes about because of the seemingly innocuous remark 'the pole will fit entirely into the barn' which we smuggled into the description above; this remark is meaningful *only* when it is associated with a particular reference frame: it is true within the farmer's wife's frame, but is simply not true within the farmer's.

In Figure 5.6, we can see the worldlines of the barn (this is another of those Minkowski diagrams which is a lot more intelligible when you build it up gradually, than when you look at the final result). Here, the front and back of the barn are labelled 'FB' and 'BB', and the front and back of the pole, which is lying along the $x'$-axis with one end at the origin, are 'FP' and 'BP'. Event ① is an event which happens at the place and time where the front of the barn and the back of the pole are coincident (the place and time of the slammed-shut door), and event ② is an event which happens at the place and time where the *back* of the barn and the *front* of the pole are coincident. We can see that events ① and ② are simultaneous in $S$ – they have the same $t$-coordinate – this is the situation where the pole is moving at a speed where

---

[5] This is also known as the 'ladder paradox', and was first described, in a slightly different form, by Rindler (1961). You can also think of it as the car-in-the-garage paradox, but do wear a seat-belt.

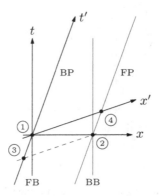

**Figure 5.6** The pole moving through the barn. Frame $S$ is the barn, frame $S'$ is attached to the pole. The worldlines FP, FB, BP, BB, are the front/back of the pole/barn. For events, see the text.

it is length contracted just the right amount to fit exactly inside the barn, in the sense of the previous paragraph. From this same diagram, you can see that event ④, which is an event which happens at the front of the pole at the same $t'$ as event ①, is beyond the back of the barn: at the instant of $t'$ when the back of the pole enters the barn in the farmer's frame (at time $t' = t'_1 = t'_4$), the front of the pole is well clear of the back of the barn, and the event when the front of the pole hits the back of the barn (event ②) is simultaneous in the farmer's frame with event ③, which is an event in front of the barn – the back of the pole hasn't entered the barn at this point. The constant spatial distance between events ① and ④, or between events ③ and ②, is the rest length of the pole, 20 m, and the distance between ① and ② is the rest length of the barn, 10 m. You can plainly see from this figure that it is impossible to find a pair of events which have the same $t'$-coordinate, which are both between the FB and BB worldlines: the pole is never entirely within the barn in the farmer's frame.

Figure 5.7 is the same set of events, but drawn in the farmer's frame, $S'$. Here, the worldlines of the barn show it to be moving in the negative $x'$-direction.

We can look at the problem another way, by asking the question 'supposing that the back of the barn were made of super-strong concrete, how would the trailing end of the pole know when to stop?' We cannot assume that the pole is completely rigid; thus when the front of the pole hits the back wall of the barn, the shock wave – the *information* that this has happened – takes a finite time to make it to the back of the pole, so that the back of the

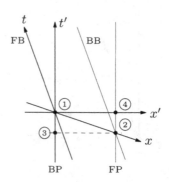

**Figure 5.7** The barn moving past the pole (this view is in the pole's frame). The events and worldlines are the same as in Figure 5.6.

pole keeps moving forwards into the barn even after the front of the pole has halted at the back wall. The information about the front of the pole stopping moves rearwards at the sound-speed in the pole: for the pole to be rigid would require an infinite sound-speed. Very shortly after the barn door has been slammed shut, the pole's recoil will smash through the door (or more likely be vaporised, but let's not worry ourselves with physical practicalities at this point).

What this example shows is that any conclusion which you correctly reach in one frame must be reachable in any other frame, even though the detailed mechanisms might be different.

Exercise 5.9, though it is framed in different language, works through this problem in illuminating detail.          [Exercise 5.9]

### 5.7.3 Bell's Accelerating Spaceships

Consider a pair of identical rockets parked on neighbouring launch pads, on the $x$-axis. They take off simultaneously (in the launch-pad frame) and fly identical flight plans, accelerating continuously along the $x$-axis.[6]

We now suppose that the rockets are connected to each other by a piece of string, which is taut when the rockets are on the launch pads. The question: does the string break during the rockets' acceleration to relativistic speeds?

Viewed in the launch-pad frame, the rockets *and the thread* are length contracted. Since the thread is length contracted, and since, in the launch-pad frame, the two spaceships remain the same distance apart from each

---

[6] This paradox is discussed in Bell (1976), and has become known as 'Bell's spaceship paradox', although he quotes it as originating somewhat earlier (Dewan & Beran 1959, Dewan 1963).

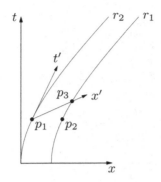

**Figure 5.8** Bell's rockets: two spaceships moving along the $x$-axis at increasingly relativistic speeds. The two rockets have worldlines $r_1$ and $r_2$.

other, the string can no longer stretch from one rocket to the other, and therefore must break at some point.

There is a contrary argument, however, which has it that, in the rockets' frame, the two rockets are stationary, so the string between them is stationary, so there's no length contraction, so the string will still stretch between the rockets, so it won't break.

Which of these arguments holds up – does the string break or not? Since the string does or does not break, one of these arguments is mistaken.

You may wish to think the arguments through before reading on. It's useful to consider how the two rockets' accelerations will be observed by the pilots of the 'other' rocket. When doing so, imagine the accelerations happening as a sequence of instantaneous increments in speed (when do these jumps happen?), rather than being continuous; also, drawing a well-chosen Minkowski diagram makes the problem very simple.

In Figure 5.8 we can see the Minkowski diagram of the rockets' motion, plus the axes of a reference frame, $S'$, which is instantaneously co-moving with the rear rocket (that is, an inertial frame which, at a specific instant, is briefly moving at the same velocity as the accelerating rocket, so that its $t'$-axis is parallel to the $r_2$ worldline at that point; $S$ and $S'$ are not in standard configuration). The key to resolving the paradox is the insight that, once the rockets are moving at speed, their clocks are no longer synchronised from each other's point of view.

Imagine one-year anniversary parties, events $p_1$ and $p_2$, on board the two spaceships, when a year of proper time has passed on board. Since the rockets have identical acceleration schedules, these will be simultaneous in the launch-pad frame (though not, of course, at $t = 1\,\mathrm{yr}$).

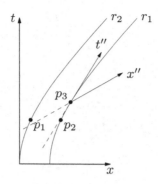

**Figure 5.9** Bell's rockets: the same as Figure 5.8, but with the axes of the front rocket's frame drawn in.

In the frame of the rear rocket, however, events $p_1$ and $p_2$ are *not* simultaneous; instead it is the events $p_1$ and $p_3$ which are simultaneous there. But event $p_3$ is after the front rocket's one-year party, and thus later in the rockets' acceleration schedule, so the rear rocket's pilot (presuming they've forgotten their relativity lectures) will conclude that the front pilot is 'cheating' on the acceleration schedule, by both organising their party ahead of time ($p_2$), and by applying more thrust than agreed, with the result that the front rocket pulls away, causing the string to break.

The piece of string, which has never attended a relativity lecture in its short and sorry life, and which is obliged to have one end at $p_1$ and the other simultaneously (in its frame[7]) at $p_3$, simply detects that the front rocket is further from the rear rocket than the string can cope with, and breaks.

If you sketch in the axes of a frame $S''$ which is instantaneously co-moving with the front rocket at $p_3$ (Figure 5.9), then you will see that the event $p_3$ is simultaneous in frame $S''$ with an event on $r_2$ with a smaller $t$ than $p_1$, that is, at an earlier point in the acceleration schedule: the front rocket's pilot (equally negligent of their relativity lectures) believes that the rear rocket has fallen behind schedule, so it is lagging behind, causing the string to break (as usual, different frames' observers have different explanations for what happens, but they must agree on the results).

The error in the contrary argument above is the phrase 'in the rockets' frame...' – there is no single 'rockets' frame' once they've started accelerating. Here, as in Figure 5.5, the key part of the resolution is to identify the

---

[7] There is no complication about 'the string's frame': even a modest continuous acceleration produces a relativistic speed before long – see Exercise 6.11 – so we can take the string to be always in equilibrium. And yes; there is such a thing as 'string theory'; no, it's not this.

frames in which things are or are not simultaneous. In Figure 5.5 Penelope 'misses' a chunk of time when she switches from frame $S'$ to $S''$; in Figure 5.8, the relativity-naïve pilots accuse each other of bad navigation, because they have not realised that both rockets have changed frame between the launch and the one-year party.                                             [Exercise 5.10]

## 5.8 Some Comments on the Lorentz Transformation

This section has a 'dangerous bend' symbol, not because it is particularly hard, but because it is something of a tangent from the thread of the text.

### 5.8.1 The Derivation of the Lorentz Transformation

The route I have chosen here, to derive the LT, is to demand that the interval Eq. (5.1) be invariant under the transformation, and then work out what transformation provides this. This is motivated by the observation, in Section 4.4, that the interval is invariant. This is partly justified by the fact that it *works*, in the sense of obtaining the LT, which is observed to match nature. But it also highlights the importance of the *geometry* of Minkowski space, which arises from our choice of $s^2$ as our definition of distance. This also leads quite naturally to the geometrical approach to General Relativity which we will meet in Appendix A.

Einstein's 1905 paper (which is reproduced in Lorentz et al. (1952)) derives the LT by what may now seem a rather circuitous route which starts with the demand that clocks in both the stationary and the moving frame are synchronised, using the procedure in Section 2.2.1, and works out what the relationship must be, between the two frames' space and time coordinates, in order for this to be so. But even there, in a footnote in §3 of that paper, he says (adjusting notation) 'The equations of the Lorentz Transformation may be more simply deduced directly from the condition that in virtue of these the relation $[\Delta x^2 = (c\Delta t)^2]$ shall have as its consequence the second relation $[\Delta x'^2 = (c\Delta t')^2]$.' The ideas of symmetry, and of frame invariance, were pervasive in Einstein's work, and have become organising principles for much of modern physics.

Taking Einstein's hint, we could choose to take the invariance of the interval as our *starting point* for relativity, instead of the second axiom of Chapter 2, and deduce from that the geometry of Minkowski space, and eventually the Lorentz transformation. This is the approach taken by Gourgoulhon (2013), for example: this is a powerful and captivating approach, but a rather abstract one; it may appeal to you if your interests run towards pure maths. Whether this approach is more or less physically illuminating is probably a matter of taste.

An alternative mathematically direct, but similarly less physically illuminating, route to the LT is to note that the transformation from the unprimed to the primed coordinates must be linear, if the equations of physics are to be invariant under a shift of origin. That is, we must have a transformation like $t' = Ax + By + Cz + Dt$, and similarly (with different coefficients) for $x'$, $y'$ and $z'$. By using the Relativity Principle and the constancy of the speed of light, one can deduce the transformation given in Eq. (5.6). There are different versions of the same discussion in Landau & Lifshitz (1975, §2), Taylor & Wheeler (1992, §§L.4–5), Barton (1999, §4.3), Rindler (2006, §2.6), and Schutz (2009, §1.6).

Finally, Rindler (2006, §2.11) shows an even more powerful consequence of the same ideas, and in particular observes that we can choose multiple alternative second axioms. If we were to choose as our second axiom 'there is no upper limit to speeds in nature', then we could examine the consequences of these axioms and before too long deduce the galilean transformation. That is, there is nothing mathematically *wrong* with this axiom – it leads to no inconsistency – it simply fails to match our universe.

If on the other hand we chose as the second axiom 'there *is* a finite upper limit to speeds in nature', then we deduce Einstein's SR (and must subsequently assert that light is a thing which travels at that upper speed). There are no options other than these two: something which moves at that upper speed-limit in one frame must be observed to move at that *same* limit speed in every other frame; if this were not the case then, by virtue of the first axiom, observers in one frame could deduce that they were moving absolutely. The second axiom could in fact be anything that is unique to SR, such as 'coordinates in inertial frames are related by the LT': the statement 'light always travels at $c$' is simply crisp and historically motivated, and shows Einstein's excellent taste in choosing a starting point for his argument.

The discussion in Section 5.8.2 shows a further alternative way of deriving the LT which, to me, is much less clear than the geometrical approach of this section. Yet again, your opinion may vary, depending on your tastes and physics background.

If you think all of these are too long-winded, you might enjoy Rindler's 'World's fastest way to get the relativistic time-dilation formula' (1967).

### 5.8.2 The Lorentz Transformation and Maxwell's Equations ⚠

It can be shown that the electric field at position $\mathbf{r} = (x, y, z)$, of a charge $q$ moving with speed $v$ along the $x$-axis, has components[8]

$$E_x = \frac{q}{4\pi\epsilon_0 r^3} x' \tag{5.15a}$$

$$E_y = \frac{q}{4\pi\epsilon_0 r^3} \gamma y \tag{5.15b}$$

$$E_z = \frac{q}{4\pi\epsilon_0 r^3} \gamma z, \tag{5.15c}$$

where $\gamma = (1 - v^2)^{-1/2}$, $r^2 = x'^2 + y^2 + z^2$, $x' = \gamma(x - \bar{x}(t))$, and $\bar{x}(t)$ is the instantaneous position of the charge at time $t$.

As is evident from Eq. (5.15), the electric field is no longer spherically symmetric, and will have a pattern which is squashed in the direction of motion, along the $x$-axis. The field around a stationary charge is symmetric, and by choosing the angular velocity carefully, we could set a charged test particle in a circular orbit around the original charge, with period $T$. It can be shown (Bell 1976) that the corresponding test particle orbiting the *moving* charge, with the squashed electric field, has an orbit compressed along the $x$-axis, with a period of $\gamma T$, greater than the period of the original system.

Bell shows that, if the motion of the particle around the moving charge is expressed using the change of variables (in units where $c = 1$)

$$x' = \gamma(x - vt) \tag{5.16a}$$

$$y' = y \tag{5.16b}$$

$$z' = z \tag{5.16c}$$

$$t' = \gamma(t - vx), \tag{5.16d}$$

then the expressions for the field of the moving charge, and the motion of the

---

[8] This expression was first established by Heaviside in 1888 or 1889 (in different notation; see Heaviside (1889)). Although the result looks reasonably simple, the calculation is famously hard: there is a discussion of the various routes to the result in Jefimenko (1994), which refers to the derivation as 'one of the most complicated procedures in classical electromagnetic theory'. The discussion in this section follows that in Bell (1976), which is also the source for 'Bell's spaceship paradox' of Section 5.7.3. Although this paper is now most commonly cited (slightly erroneously) merely as the source of the paradox, Bell's goal in the paper was to claim that the argument in this section is a more intuitive way of arriving at SR than the axiomatic approach. I think this is true only for those with a substantial pre-existing familiarity with advanced EM theory, but that it is largely unintelligible for those without this familiarity, as well as missing the fundamental insights mentioned at the end of this section.

test particle in orbit around it, reduce to the same form as the (spherically symmetrical) expressions for a stationary charge, and the particle again orbits with period $T$; and further, that Maxwell's equations in these changed variables have the same form as Maxwell's equations in the original variables (compare Exercise 2.4).

Equations (5.16) are of course now familiar. These and similar expressions were well known to physicists for at least a decade prior to Einstein's 1905 paper. The LT is also sometimes referred to as the FitzGerald–Lorentz transformation, since it appears to have been George Francis FitzGerald who first suggested (1889) that 'the length of material bodies changes, according as they are moving through the ether or across it', which was elaborated as an idea by Hendrik Antoon Lorentz (1895, 1904). It may be J. J. Thomson who first put the change of variables Eq. (5.16a) into print (Thomson 1889), as a way of recovering the spherical symmetry of Maxwell's equations, in the case of a particle in motion in an electric field, but it does seem that it was Lorentz who came closest to writing down, what it now seems fair to call, the Lorentz transformation. The papers I have mentioned here form part of a network of dauntingly complicated attempts to discuss the consequences of Maxwell's equations in a frame moving with respect to the aether and – although this is certainly not how these authors would have described their work – to find a transformation which left form-invariant the equations of electromagnetism. It's easy to think, now, of SR as part of mechanics (think of all these trains moving about at high speed and, in chapters to come, particles colliding), and to be slightly puzzled at why it would occur to Einstein to think of specifically *light* as having some special status. But Einstein's 1905 paper is, even looking only at its title, located within the study of electromagnetism; and almost half of it, subtitled 'Electrodynamical part' and beginning with 'Transformation of the Maxwell–Hertz equations for empty space', is manifestly connected to the same network of concerns as above. Despite thus advertising its connections with previous work, the 1905 paper is remarkable for the completeness and concision of its reboot of physics as the study of symmetry in nature, which Minkowski could be said to declare as complete in 1908.

Recall that, at this point in this section, we have *not* yet talked about Special Relativity; so far, this is all (advanced) classical electromagnetism.

If the 'moving charge' described above is the nucleus of an atom, and the orbiting charges are the atomic electrons, then the above analysis is telling us that those atoms will be 'squashed' along the direction of motion by a factor of $\gamma$, and thus that a rod which is, for example, Avogadro's number of atoms long will be contracted in length by the same factor. If the 'rod' is instead your brain, in motion while sitting in a travelling train, then not only will it be contracted along the direction of your motion, but all of the electrons around all of the atoms in it will be orbiting with a period which is longer than the period at rest, by another factor of $\gamma$. The result, presumably, would be that you would think slower by a factor of $\gamma$, and the watch in your hand – with its components also made of atoms – would tick slower by a factor of $\gamma$, ticking out $t'$ rather than $t$. Thus all of your observations within the carriage would be in terms of $(t', x')$ rather than the $(t, x)$ appropriate to measurements on the stationary platform. But that in turn means that all of the electromagnetic measurements you make, in the carriage, would be indistinguishable from the corresponding measurements made on the platform, even though they would be measured to go more slowly by observers standing on that platform. It would be impossible, based on experiments and observations made within the carriage, for you to discover whether you were in motion.

It is possible to imagine, at this point, a counterfactual history of relativity without Einstein, in which all of the equations of SR are obtained as consequences of Eq. (5.16), and we discover, as a consequence of these, the invariance of the interval $s^2$. The measurable results would be largely the same as what we have now, but the route and the foundations would be very different.

We might end up concluding that an absolute rest frame was a physically real thing, even though, by a peculiarity of Maxwell's equations, it was impossible to detect (this appears to have been Lorentz's position, just as, looking further back, it may have been Newton's position even though his theory, similarly, contained nothing which would allow the rest frame to be identified). If you believe in atoms, and in particular atoms composed of electrons orbiting a nucleus (a belief which was not universal in the first decades of the twentieth century), then this feature of Maxwell's equations transfers itself to material bodies, and thus to the rest of physics. The argument above, about the atoms in your brain, makes it *plausible* that you would think slower whilst moving, but it doesn't prove it. All together, this argument from Maxwell's equations explains why we don't see an aether and why 'moving clocks run slow', but leaves a large number of questions still open, and in addition doesn't address the question of *why* it should be that Maxwell's equations have this odd Lorentz-group symmetry, rather than anything more intuitive.

Einstein's axiomatic starting point, in contrast, focuses on the underlying geometry of, and properties of, spacetime, rather than one particular physical theory, and it therefore immediately has universal applicability.

Einstein's two axioms say nothing of electromagnetism, nor of mechanics: the first axiom says that all inertial frames are equivalent for *all physical experiments* (the absolute in Einstein's work is not absolute rest, but absolute symmetry). This axiom is consistent with the galilean transformation as well as Lorentz's, so the role of the second axiom is to pick out the latter from the former: saying 'light always moves at $c$' is a fine second axiom, saying '$\Delta t^2 - \Delta x^2$ is frame-invariant' would be a perfectly adequate alternative (as Einstein hints in a footnote in the 1905 paper, and as I discussed in Section 5.8.1).

At the risk of taking a tangent to the tangent, this seems a good place to observe that the 'theory of *relativity*' is something of a misnomer, since calling it relativity muddles the problem (the frame dependence of coordinates) with the solution (the invariance of the interval).[9] That is, Einstein's theory solves the problem by identifying new absolutes, most specifically the invariance of $s^2$ between frames. In the nearest that Einstein came to an autobiography (Einstein 1991), he repeatedly stresses the methodological and philosophical importance of identifying invariance or symmetry under transformation.

## Exercises

### Exercise 5.1 (§5.1)

Look back at Figure 3.3 on p. 34, and consider the following two frames. Frame $S$ is attached to the right-moving carriage and has its spatial origin $x = 0$ at the centre of the top carriage; frame $S'$ is attached to the left-moving carriage and has its spatial origin $x' = 0$ at the centre of that carriage. Sketch the position of these carriages/frames at time $t = 0$. Are these frames in standard configuration? Can we use the Lorentz transformation to relate the coordinates of these frames?          $[u^+]$

---

[9] It's a fairly common observation that Einstein disliked the term, but it's unexpectedly hard to find a specific source for it. The nearest I have been able to find is a 1921 letter to Eberhard Zschwimmer (Einstein Papers, volume 12, document 250, https://einsteinpapers.press.princeton.edu/vol12-doc/372), in which he says that the name 'relativity theory' can lead to philosophical misunderstandings, and agrees that 'invariance theory' (*Invariantz-Theorie*) would better describe the method; he acknowledges, however, that it was even then too late to change the term.

## Exercise 5.2 (§5.1)

A spaceship cruises past the Earth travelling at speed $v = \sqrt{3}/2$. The ship is carrying a beacon flashing 120 times per minute (that is, $0.5\,\text{s}$ between flashes).

Explain how you would use a network of observers to make direct measurements of the length of the spaceship, and the interval between flashes, in the Earth's frame (that is, without using the LT).

Use the time dilation formula to obtain the interval between flashes, as measured on Earth.

Earth traffic control sends a radio signal to the ship $10^{10}\,\text{m}$ (a little under a minute) after the ship passes, as measured on Earth; the flashing stops at a time $3 \times 10^{10}\,\text{m}$ after the flyby, as measured on the ship. Sketch these two events on a Minkowski diagram, obtain their coordinates, and calculate the invariant interval between them. State whether the end of the flashing could be attributed to the signal from traffic control, and why.

## Exercise 5.3 (§5.2)

What happens to $\gamma$ as we take the velocity $v$ towards zero in Eq. (5.7) – that is, as we move to the non-relativistic velocities of our everyday experience?

The LT of Eq. (5.6), written in physical units, is

$$ct' = \gamma\left(ct - \frac{vx}{c}\right)$$
$$x' = \gamma(x - vt).$$

What is the form of the LT in this slow-speed limit? Do you recognise this?

From Eq. (5.11), how do velocities add in this limit, where either $v_1$ or $v_2$ is small compared with $c = 1$? And what happens if one of the velocities is already the speed of light?

## Exercise 5.4 (§5.2)

In the notation of Eq. (5.10), show that $\gamma(v) = \cosh\phi$, and that

$$\frac{\gamma(v)}{\gamma(v_1)\gamma(v_2)} = 1 + v_1 v_2.$$

## Exercise 5.5 (§5.5.1)

In Section 5.5.1 the LT was used to obtain the length-contraction formula (Eq. (5.14)) directly. Use a similar argument to obtain the time-dilation result of Eq. (3.3), using the LT of Eq. (5.6).

## Exercise 5.6 (§5.5.2)

What is the speed of the trains in Figure 3.2 and Figure 3.3 (p. 34)? Remember that the train carriage is 6 m long, and that the clocks are showing times in units of metres.

Further, show that the times shown in Figure 3.5 are correct.

What is the length of the lower train, which Barbara and Hilary measure (i.e., $L'$ in Figure 3.4)? [This question is a little tricky, not because it's conceptually hard, but because it's a bit like a logic puzzle, in seeming to have too little information; getting the right choice of frames is important.]

$$[d^+]$$

## Exercise 5.7 (§5.5.2)

A train enters a tunnel travelling at a speed $3/5$ (in units where $c = 1$). The tunnel is 500 m long, and the train 100 m long, both as measured in their rest frames. The train has clocks at the extreme front and back, which are synchronised in the train's frame. The rear clock is observed to show time 0 m at the instant the rear of the train disappears into the tunnel. The front clock is also observed when it emerges from the other end of the tunnel.

Draw a Minkowski diagram for the frame of the tunnel, showing the train's motion, including the worldlines of the clocks and the tunnel ends, and the three events consisting of ① the rear clock disappearing into the tunnel, an event ② at the front of the train, simultaneous with ① in $S'$, and an event ③ at the point where the front clock emerges from the tunnel.

By using the Lorentz transformation and/or the length-contraction formula, or otherwise, deduce the time on the front clock when it emerges from the tunnel.

## Exercise 5.8 (§5.7.1)

In Figure 5.10, we see two alternative routes between events ① and ③, namely a direct route, and one via a second event ②. Take the interval between events ② and ③ to be null. By considering the spatial, temporal, and invariant intervals between these three events, work out the total invariant

**Figure 5.10** A 'dog-leg' path in Minkowski space.

interval along the two paths, and thus conclude that in Minkowski space a straight line is the *longest* timelike interval between two points.

It is reasonable to assume that the total interval along a path is the sum of the intervals along its segments.                    [ $d^+$ ]

## Exercise 5.9 (§5.7.2)

Take two cars, moving along the $x$-axis at a speed $v$, separated by 1 km in their frame. The lead car passes first a checkpoint and then, 0.5 km further on in the checkpoint frame, a traffic policeman.

Sketch a Minkowski diagram of these events, indicating at least the world-lines of the cars, the checkpoint and the policeman, and marking the events:

- ①, the lead car passing the checkpoint;
- ②, the position of the second car at the same time as ① in the cars' frame;
- ③, the position of the second car at the same time as ① in the checkpoint frame;
- ④, the second car reaching the checkpoint; and
- ⑤, the first car reaching the policeman.

[It will probably help if you choose the frames such that event ① has coordinates $x_1 = t_1 = x'_1 = t'_1 = 0$, but if you want to do it another way, that's fine.]

Give expressions for the coordinates of these events (in either $S$ or $S'$ as appropriate), as expressions in terms of the speed $v$. This is the 'write down what you know' step of the Appendix D recipe.

Give an expression for the interval $s^2_{45}$ between events ④ and ⑤, in terms of $v$, and calculate numerical values for this in the cases (i) $v = 1/2$, (ii) $v = 3/5$ and (iii) $v = 4/5$. In each of the three cases, state, with an explanation, whether it is possible for the traffic policeman to signal to the checkpoint to lower a barrier before the second car arrives.

[Optional extra: draw Minkowski diagrams for the three cases (i), (ii) and (iii), which illustrate the answers you obtain above. This might illuminate both the answer to this question, and the pole-in-the-barn problem in the farmer's frame.]                                                          $[u^+]$

## Exercise 5.10 (§5.7.3)

A piezoelectric crystal is one which generates an electrical voltage when it is squashed. Imagine that the 'carriages' in Figure 3.2 were in fact piezoelectric crystals, and recall that we worked out in Section 3.2 that the observers on the other carriages (and, though we did not work this out explicitly, the observers on the station platform) would measure the top carriage/crystal to be shorter than it would be when it was at rest. Would the crystal be generating a voltage because it's shorter?                                     $[d^+]$

# 6

# Vectors and Kinematics

In Chapter 5, we used the axioms of Chapter 2 to obtain the Lorentz transformation. That allowed us to describe events in two different frames in relative motion. That part was rather mathematical in style. Now we are going to return to the physics, and describe *motion*: velocity, acceleration, momentum, energy and mass.

*Aims*: you should:

6.1. understand the concept of a 4-vector as a geometrical object, and the distinction between a vector and its components.

Chapter 5 was concerned with static events as observed from moving frames. In this part, we are concerned with particle motion.

Before we can explain motion, we must first be able to describe it. This is the subject of *kinematics*. We will first have to define the vectors of four-dimensional Minkowski space, and specifically the velocity and acceleration vectors.

## 6.1 Three-Vectors

You are familiar with 3-vectors – the vectors of ordinary three-dimensional euclidean space. To an extent, 3-vectors are merely an ordered triple of numbers, but they are interesting to us as physicists because they represent a more fundamental geometrical object: the three numbers are not just picked at random, but are the vector's *components* – the projections of the vector onto three *orthogonal* axes (that the axes are orthogonal is not essential to the definition of a vector, but it is almost always simpler than the alternative). That is, the components of a vector are functions of both the vector and our

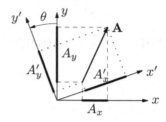

**Figure 6.1** A displacement vector in 3-d euclidean space.

choice of axes, and if we change the axes, then the components will change in a systematic way.

For example, consider a prototype displacement vector $(\Delta x, \Delta y, \Delta z)$. These are the components of a vector with respect to the usual axes $\mathbf{e}_x$, $\mathbf{e}_y$ and $\mathbf{e}_z$. If we rotate these axes, say by an angle $\theta$ about the $z$-axis, to obtain axes $\mathbf{e}'_x$, $\mathbf{e}'_y$ and $\mathbf{e}'_z$, we obtain a new set of coordinates $(\Delta x', \Delta y', \Delta z')$, related to the original coordinates by

$$
\begin{pmatrix} \Delta x' \\ \Delta y' \\ \Delta z' \end{pmatrix} = \begin{pmatrix} \cos\theta & \sin\theta & 0 \\ -\sin\theta & \cos\theta & 0 \\ 0 & 0 & 1 \end{pmatrix} \begin{pmatrix} \Delta x \\ \Delta y \\ \Delta z \end{pmatrix}. \tag{6.1}
$$

These new components describe the *same* underlying vector $\mathbf{A} = (\Delta x, \Delta y, \Delta z)$, as shown in Figure 6.1 (omitting the $z$-axis). Although any random triple of numbers $(\Delta x, \Delta y, \Delta z)$ describes *some* vector, to some extent what turns the number triple into a *vector* is the existence of this underlying object, which implies all the other sets of coordinates $(A'_x, A'_y, A'_z) = (\Delta x', \Delta y', \Delta z')$, that have a particular functional relation to the original $(A_x, A_y, A_z) = (\Delta x, \Delta y, \Delta z)$.

Note that this prototype vector is a *displacement vector*; the 'position vector', in contrast, which goes from the origin to a point, is *not* a vector in this sense.

Ideally, you should think of the vectors here as being defined as a length plus direction, defined at a point, rather than an arrow spread out over a finite amount of space (you might think of this as some sort of infinitesimal displacement vector). If you think of an electric or magnetic field, which has a size and a direction at every point, you'll have a very good picture of a vector in the sense we're using it here.

There is a swift review of linear algebra in Section C.3; make sure you are comfortable with the terms and definitions mentioned there.

## 6.2 Four-Vectors

As we saw in Section 5.3, we can regard the events of SR taking place in a 4-dimensional space termed spacetime. Here, the prototype displacement 4-vector is $(\Delta t, \Delta x, \Delta y, \Delta z)$, relative to the space axes and wristwatch of a specific observer, and the transformation which takes one 4-vector into another is the familiar LT of Eq. (5.6), or

$$\begin{pmatrix} \Delta t' \\ \Delta x' \\ \Delta y' \\ \Delta z' \end{pmatrix} = \begin{pmatrix} \gamma & -\gamma v & 0 & 0 \\ -\gamma v & \gamma & 0 & 0 \\ 0 & 0 & 1 & 0 \\ 0 & 0 & 0 & 1 \end{pmatrix} \begin{pmatrix} \Delta t \\ \Delta x \\ \Delta y \\ \Delta z \end{pmatrix} \tag{6.2a}$$

for the 'forward transformation' and

$$\begin{pmatrix} \Delta t \\ \Delta x \\ \Delta y \\ \Delta z \end{pmatrix} = \begin{pmatrix} \gamma & +\gamma v & 0 & 0 \\ +\gamma v & \gamma & 0 & 0 \\ 0 & 0 & 1 & 0 \\ 0 & 0 & 0 & 1 \end{pmatrix} \begin{pmatrix} \Delta t' \\ \Delta x' \\ \Delta y' \\ \Delta z' \end{pmatrix} \tag{6.2b}$$

for the inverse transformation (that the matrices are inverses of each other can be verified by direct multiplication). These give the coordinates of the same displacement as viewed by a second observer whose frame is in standard configuration with respect to the first.

This displacement 4-vector $\Delta\mathbf{R} = (\Delta t, \Delta x, \Delta y, \Delta z)$ we can take as the prototype 4-vector, and recognise as a 4-vector anything which transforms in the same way under the coordinate transformation of Eq. (6.2a) (this may seem a rather abstract way of defining vectors, but we will see a concrete example in Section 6.5).

We write the components of a general vector as $\mathbf{A} = (A^0, A^1, A^2, A^3)$ (do note that the superscripts are indexes, not powers),[1] or collectively $A^\mu$, where the greek index $\mu$ runs from 0 to 3. We will also occasionally use latin superscripts like $i$ or $j$: these should be taken to run over the 'space' indexes, from 1 to 3.

An arbitrary vector $\mathbf{A}$ has *components* $A^\mu$ in a frame $S$, as illustrated in Figure 6.2. The components $A^1$, $A^2$ and $A^3$ (that is, $A^i$) are just as you would expect, namely the projections of the vector $\mathbf{A}$ onto the $x$-, $y$- and $z$-axes. The component $A^0$ is the projection of the vector onto the $t$-axis – it's the *timelike* component. In a frame $S'$, the space and time axes will be different,

---

[1] This is now, I think, the most common notation, but you can still find books which observe the slightly more old-fashioned convention of labelling these as $(A^1, A^2, A^3, A^4)$.

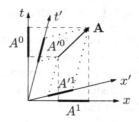

**Figure 6.2** A displacement vector in 4-d Minkowski space.

and so the projections of the vector **A** onto these axes will be different. Just as in Figure 6.1, we find the *projection* of a point by moving it parallel to one of the axes. For example, we find the projection onto the $x$-axis of the end of the vector **A**, by moving that point parallel to the $t$-axis until it hits the $x$-axis, and we find the projection of the same point onto the $x'$-axis by moving it parallel to the $t'$-axis until it hits the $x'$-axis.

Given the components of the vector in one frame, we want to be able to work out the components of the *same* vector in another frame. We can obtain this transformation by direct analogy with the transformation of the displacement 4-vector, and take an arbitrary vector **A** to transform in the same way as the prototype 4-vector $\Delta$**R**.

That is, given an arbitrary vector **A**, the transformation of its components $A^\mu$ in $S$ into its components $A'^\mu$ in $S'$ is exactly as given in Eq. (6.2a):

$$\begin{pmatrix} A^{0'} \\ A^{1'} \\ A^{2'} \\ A^{3'} \end{pmatrix} = \begin{pmatrix} \gamma & -\gamma v & 0 & 0 \\ -\gamma v & \gamma & 0 & 0 \\ 0 & 0 & 1 & 0 \\ 0 & 0 & 0 & 1 \end{pmatrix} \begin{pmatrix} A^0 \\ A^1 \\ A^2 \\ A^3 \end{pmatrix}. \tag{6.3a}$$

Since this is a matrix equation, the inverse transformation is straightforward: it is just the matrix inverse of this:

$$\begin{pmatrix} A^0 \\ A^1 \\ A^2 \\ A^3 \end{pmatrix} = \begin{pmatrix} \gamma & \gamma v & 0 & 0 \\ \gamma v & \gamma & 0 & 0 \\ 0 & 0 & 1 & 0 \\ 0 & 0 & 0 & 1 \end{pmatrix} \begin{pmatrix} A^{0'} \\ A^{1'} \\ A^{2'} \\ A^{3'} \end{pmatrix}. \tag{6.3b}$$

You may also see this written out in matrix form as

$$\mathbf{A'} = \Lambda\mathbf{A}; \qquad \mathbf{A} = \Lambda^{-1}\mathbf{A'} \tag{6.4}$$

where the transformation matrix $\Lambda$ is as in Eq. (6.3a). Written out explicitly,

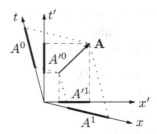

**Figure 6.3** The displacement vector of Figure 6.2 in the 'other' frame.

the expressions in Eq. (6.3a) are

$$
\begin{aligned}
A^{0'} &= \gamma(A^0 - vA^1) \\
A^{1'} &= \gamma(A^1 - vA^0) \\
A^{2'} &= A^2 \\
A^{3'} &= A^3,
\end{aligned}
\tag{6.5}
$$

and, as you can see, these have the same form as the more familiar expression for the Lorentz transformation of an event's coordinates, as in Eq. (5.6). Note also that the primed and unprimed frames are perfectly symmetrical. The observers at rest in the unprimed frame $S$ regard the primed frame $S'$ as moving at speed $v$ with respect to them. However, the observers in the primed frame see the unprimed one moving at speed $-v$; they would also naturally swap the assignment of primed and unprimed frame, with their frame, say $P$, being the unprimed one and the other, $P'$, being the primed one. In Figure 6.3 we can see the vector **A** in Minkowski space, drawn from the point of view of these 'primed' observers.

Now, these two sets of observers are describing the *same* events and frames, but with opposite notation, and so with an opposite sign for $v$. We can translate between the two sets of notations by simultaneously swapping primed and unprimed quantities and swapping the sign of $v$. But all this is, is a way of translating between a transformation and its inverse. This is exactly what happens between, for example, Eq. (5.6) and Eq. (5.8), or between Eq. (6.2a) and Eq. (6.2b).

I've talked repeatedly of events being frame-independent. Vectors are *also* frame-independent, *even though their components are not* (this is what distinguishes a vector from being just a collection of four numbers).

More linear algebra, directly echoing the corresponding results in the previous section, and in Section C.3: Just as with 3-vectors, if **A** and **B** are 4-vectors, so is $a\mathbf{A}$ (for all $a \in \mathbb{R}$), and so is $\mathbf{A} + \mathbf{B}$, with components

$(A^0 + B^0, A^1 + B^1, A^2 + B^2, A^3 + B^3)$. Where the scalar product of ordinary 3-vectors is straightforward, the different geometry of spacetime means that the useful scalar product for 4-vectors is defined as

$$\mathbf{A} \cdot \mathbf{B} = A^0 B^0 - A^1 B^1 - A^2 B^2 - A^3 B^3 \tag{6.6a}$$

$$= \sum_{\mu,\nu} \eta_{\mu\nu} A^\mu B^\nu \tag{6.6b}$$

$$= \mathbf{A}^T \eta \mathbf{B}, \tag{6.6c}$$

where the matrix $\eta_{\mu\nu}$ is defined as

$$\eta_{\mu\nu} = \mathrm{diag}(1, -1, -1, -1) \tag{6.7}$$

(denoting the matrix with components which are all zero except for $(1, -1, -1, -1)$ on its diagonal), and in the last line I have written the equivalent matrix expression involving column vectors $\mathbf{A}$ and $\mathbf{B}$. The scalar product of a vector with itself, $\mathbf{A} \cdot \mathbf{A}$, is its *magnitude* (or length-squared, written $|\mathbf{A}|^2$ or $A^2$), and from this definition we can see that the magnitude of the displacement vector is $\Delta \mathbf{R} \cdot \Delta \mathbf{R} = \Delta t^2 - \Delta x^2 - \Delta y^2 - \Delta z^2 = s^2$.

This matrix $\eta_{\mu\nu}$ is known as the *metric*, and through its use in defining the magnitude of the vector, it *defines* what 'distance' means in Minkowski space; that is, we could take the above definition of $\Delta \mathbf{R} \cdot \Delta \mathbf{R}$ as the definition of the invariant interval $s^2$ (we will see a lot more of this when we look at General Relativity). You can verify by explicit calculation that the matrix $\eta$ has the property that, when it is acted upon by the transformation matrix of Eq. (6.3a),

$$\Lambda^T \eta \Lambda = \eta \tag{6.8}$$

(where $\Lambda^T$ is the transpose of the matrix $\Lambda$). That is, it transforms into itself, telling us that the definition of distance in one coordinate system is the same in every transformed coordinate system; this is another statement of the frame-independence of the invariant interval, $s^2$.

The metric of three-dimensional euclidean space is $\mathrm{diag}(1, 1, 1)$, which (for 3-vector $\mathbf{a}$, and by analogy with Eq. (6.6)) gives us $\mathbf{a} \cdot \mathbf{a} = a_x^2 + a_y^2 + a_z^2$, or Pythagoras's theorem, as the definition of distance in euclidean space.

Just as we discussed in Section 4.7 for intervals, 4-vectors can be timelike, spacelike or null, depending on whether their magnitude is positive, negative or zero; note that, since the magnitude is not positive-definite (i.e., it can be negative), even a non-zero vector can be null. Just as in Section 6.1, we say that two vectors are *orthogonal* if their scalar product vanishes. For example, the two vectors $A = (1, 2, 0, 0)$ and $B = (2, 1, 0, 0)$ have scalar product $1 \times$

$2 - 2 \times 1 = 0$, and so are orthogonal. It follows that, in this geometry, a null vector (with $\mathbf{A} \cdot \mathbf{A} = 0$) is orthogonal to itself!

The magnitude of a 4-vector is frame-invariant. The magnitude of $\mathbf{A} + \mathbf{B}$ is $(\mathbf{A} + \mathbf{B}) \cdot (\mathbf{A} + \mathbf{B}) = A^2 + B^2 + 2\mathbf{A} \cdot \mathbf{B}$, and since $|\mathbf{A} + \mathbf{B}|^2$, $A^2$, and $B^2$ are all three frame-invariant, the scalar product $\mathbf{A} \cdot \mathbf{B}$ must be frame-invariant also.

We can see this explicitly by calculating the scalar product of $\mathbf{A}' = \Lambda\mathbf{A}$ and $\mathbf{B}' = \Lambda\mathbf{B}$:

$$\begin{aligned}
\mathbf{A}' \cdot \mathbf{B}' &= (\Lambda\mathbf{A}) \cdot (\Lambda\mathbf{B}) && \text{using Eq. (6.4)} \\
&= (\Lambda\mathbf{A})^T \eta (\Lambda\mathbf{B}) && \text{using Eq. (6.6)} \\
&= \mathbf{A}^T (\Lambda^T \eta \Lambda)\mathbf{B} && \\
&= \mathbf{A}^T \eta \mathbf{B} && \text{using Eq. (6.8)} \\
&= \mathbf{A} \cdot \mathbf{B}.
\end{aligned}$$

[Exercises 6.1–6.4]

### 6.2.1 Some Observations on 4-vectors

There's a lot going on in this section, and a lot of rather profound maths is being glanced at, to do with symmetry, invariance, and some of the mathematical physics that underlies General Relativity. None of this is necessary to understand the sections that follow, but the rather miscellaneous remarks below address some questions of detail which may occur to those with a mathematical background.

It's no coincidence that the same expression appears in both Eq. (5.6) and Eq. (6.5), but it's important that you realise that slightly different things are happening in the two cases. In Eq. (5.6), the LT relates the *coordinates of a single event* in two reference frames; in Eq. (6.3), the LT relates the *components of a single vector* in two frames. These are not the same thing, because *the coordinates of an event are not the components of a 'position vector'*. The 'position vector' is not a vector in the terms of this section, because the position vector is not frame independent; since it stretches from the frame origin to the position of something, it is obviously different in different frames. This is why it is the *displacement* 4-vector that we take as the prototype 4-vector above; since the events (i.e., locations in spacetime) at each end of the displacement vector are frame-independent, the displacement 4-vector, taken as a geometrical object, is frame-independent also.

There's an element of arbitrariness in the choice of the matrices in Eq. (6.1) and Eq. (6.2a), but these choices have the crucial property of preserving the magnitude of the appropriate displacement vector (i.e., preserving euclidean or lorentzian geometry). Similarly, the definition of $\eta$ in Eq. (6.7) provides a scalar product which is physically meaningful.

The set of numbers $\eta_{\mu\nu}$ are not just any old matrix, but in particular the components of a *tensor* – a mathematical structure which is a generalisation of a vector, and which is vital for the mathematical study of General Relativity. It is not even an arbitrary tensor: it is the form taken in a non-accelerating frame (ie, in SR) by the *metric tensor* which is a crucial object in General Relativity. It is the metric tensor which encapsulates the notion of the *distance between* two events in spacetime.

Related to that, the positions of the indexes in Eq. (6.6) are *not* arbitrary. The details do not matter for us here, although they are crucial in a study of General Relativity, but I have preserved the notational convention in order to remain at least visually compatible with any more advanced texts you may consult.

While we're on the subject of mathematical niceties, I'll observe that, since the magnitude of a vector can be negative, we can't in general take its square root to find the quantity that we might want to call the 'norm' or length of the vector; this is why the word 'distance' is in scare-quotes in this section. Secondly, the scalar or 'dot' product is the only vector product in Minkowski 4-space; there is no cross-product defined (that's a peculiarity of euclidean 3-space). Finally, and firmly in the tradition of physicists being rather casual about mathematical niceties, you will sometimes see $\mathbf{A} \cdot \mathbf{B}$ being referred to as an 'inner product'; a mathematician would insist that an inner product must be positive semidefinite, which Eq. (6.6) indicates is not true for the Minkowski scalar product.

> 'When *I* use a word,' Humpty Dumpty said in rather a scornful tone, 'it means just what I choose it to mean – neither more nor less.'
> 'The question is,' said Alice, 'whether you *can* make words mean so many different things.'
> 'The question is,' said Humpty Dumpty, 'which is to be master – that's all.'
> Lewis Carroll, *Through the Looking Glass*

[Exercise 6.5]

## 6.3  Velocity and Acceleration

Since the displacement 4-vector $\Delta\mathbf{R}$ is a vector (in the sense that it transforms properly according to Eq. (6.2a)), so also is the infinitesimal displacement $d\mathbf{R}$. We can write the components of $d\mathbf{R}$ as $dx^{\mu}$, so that the magnitude of this infinitesimal displacement vector is $d\mathbf{R} \cdot d\mathbf{R} = (dx^0)^2 - (dx^1)^2 - (dx^2)^2 - (dx^3)^2$, or $dt^2 - |d\mathbf{r}|^2$ (writing $d\mathbf{r}^2 = dx^2 + dy^2 + dz^2$). Since the proper time $\tau$ (see Section 5.4) is a Lorentz *scalar*,[2] we can divide each component of this infinitesimal displacement by the proper time and still have a vector. This latter vector is the *4-velocity*:

$$\mathbf{U} = \frac{d\mathbf{R}}{d\tau} = \left(\frac{dx^0}{d\tau}, \frac{dx^1}{d\tau}, \frac{dx^2}{d\tau}, \frac{dx^3}{d\tau}\right), \tag{6.9}$$

By the same argument, the 4-acceleration

$$\mathbf{A} = \left(\frac{d^2x^0}{d\tau^2}, \frac{d^2x^1}{d\tau^2}, \frac{d^2x^2}{d\tau^2}, \frac{d^2x^3}{d\tau^2}\right) \tag{6.10}$$

is a 4-vector, also. We can more naturally write these as

$$U^\mu = \frac{dx^\mu}{d\tau}$$

and

$$A^\mu = \frac{dU^\mu}{d\tau} = \frac{d^2x^\mu}{d\tau^2}.$$

Let us examine these components in more detail, taking $d\mathbf{R}$ to be the infinitesimal displacement of a particle. In a frame (a coordinate system) in which the particle is not moving, $d\mathbf{r} = 0$, so that $d\mathbf{R} \cdot d\mathbf{R} = d\tau^2$, where $\tau$ is the proper time (remember that proper time is a 'clock attached to the particle'). Since this magnitude of $d\mathbf{R}$ is frame-invariant,

$$d\tau^2 = dt^2 - |d\mathbf{r}|^2,$$

so that

$$\left(\frac{d\tau}{dt}\right)^2 = \frac{(d\tau)^2}{(dt)^2} = 1 - \frac{|d\mathbf{r}|^2}{(dt)^2} = 1 - v^2 = \frac{1}{\gamma^2}.$$

Taking the square root and inverting, we immediately find that

$$\frac{dt}{d\tau} = \gamma, \tag{6.11}$$

---

[2] By 'scalar' I merely mean 'a number', in a context where we want to distinguish the quantity from a vector. Note that a number is necessarily frame-invariant.

and so

$$U^0 = \frac{\mathrm{d}x^0}{\mathrm{d}\tau} = \frac{\mathrm{d}t}{\mathrm{d}\tau} = \gamma \qquad (6.12a)$$

$$U^i = \frac{\mathrm{d}x^i}{\mathrm{d}\tau} = \frac{\mathrm{d}t}{\mathrm{d}\tau}\frac{\mathrm{d}x^i}{\mathrm{d}t} = \gamma v^i. \qquad (6.12b)$$

You can view Eq. (6.11) as yet another manifestation of time dilation. Thus we can write

$$\mathbf{U} \equiv (U^0, U^1, U^2, U^3) = (\gamma, \gamma v^x, \gamma v^y, \gamma v^z) = \gamma(1, v^x, v^y, v^z). \qquad (6.13a)$$

We will generally write this, below, as

$$\mathbf{U} = \gamma(1, \mathbf{v}), \qquad (6.13b)$$

using $\mathbf{v}$ to represent the three (space) components of the (spatial) velocity vector, but as notation this is perhaps a little 'slangy'.

Note that I will consistently use upper-case letters for 4-vectors (either as a vector $\mathbf{A}$, written with bold-face, or referring to their components $A^\mu$), and lower-case letters for 3-vectors (either $\mathbf{a}$ or $a^i$).

In a frame which is co-moving with a particle, the particle's velocity is $\mathbf{U} = (1,0,0,0)$, so that, from Eq. (6.6), $\mathbf{U} \cdot \mathbf{U} = 1$; since the scalar product is frame-invariant, it must have this same value in *all* frames, so that, quite generally, we have the relation

$$\mathbf{U} \cdot \mathbf{U} = 1. \qquad (6.14)$$

You can confirm that this is indeed true by applying Eq. (6.6) to Eq. (6.13).

Here, we defined the 4-velocity by differentiating the displacement 4-vector, and deduced its value in a frame co-moving with a particle. We can now turn this on its head, and *define* the 4-velocity as a vector which has magnitude 1 and which points along the $t$-axis of a co-moving frame (this is known as a 'tangent vector', and is effectively a vector 'pointing along' the worldline). We have thus defined the 4-velocity of a particle as the vector which has components $(1, \mathbf{0})$ in the particle's rest frame. Note that the magnitude of the vector is always the same; the particle's speed relative to a frame $S$ is indicated not by the 'length' of the velocity – its magnitude, which is always 1 – but by the *direction* of the vector in Minkowski space, in the frame $S$. We can then *deduce* the form in Eq. (6.13) as the Lorentz-transformed version of $(1, \mathbf{0})$. Compare Section 6.4 and Section 7.1.1.

Equations (6.12) can lead us to some intuition about what the velocity vector is telling us. When we say that the velocity vector in the particle's rest frame is $(1, \mathbf{0})$, we are saying that, for each unit proper time $\tau$, the particle moves the same amount through coordinate time $t$, and not at all through space $x$; the particle 'moves into the future' directly along the $t$-axis. When we are talking instead about a particle which is moving with respect to some frame, the equation $U^0 = dt/d\tau = \gamma$ tells us that the particle moves through a greater amount of this frame's coordinate time, $t$, per unit proper time (where, again, the 'proper time' is the time showing on a clock attached to the particle) – 'time dilation' yet again.

Turning now to the acceleration $\mathbf{A}^\mu$, we have

$$A^0 = \frac{d^2 x^0}{d\tau^2} = \frac{dU^0}{d\tau} = \frac{d\gamma}{d\tau} = \gamma \frac{d\gamma}{dt} \equiv \gamma\dot{\gamma} \tag{6.15a}$$

$$A^i = \frac{d^2 x^i}{d\tau^2} = \frac{dU^i}{d\tau} = \gamma \frac{d}{dt}(\gamma v^i) = \gamma\left(\dot{\gamma} v^i + \gamma a^i\right) \tag{6.15b}$$

(where $v^i$ and $a^i$ are the ordinary velocity and acceleration $v^i = dx^i/dt$, $a^i = d^2 x^i/dt^2$, and $\dot{\gamma} = d\gamma/dt$), or

$$\mathbf{A} = \gamma\left(\dot{\gamma}, \dot{\gamma}\mathbf{v} + \gamma\mathbf{a}\right). \tag{6.16}$$

There is no inertial frame in which an accelerating particle is always at rest; at any instant, however, such a particle has a definite velocity, and so there is a frame – the *instantaneously co-moving* inertial frame (ICRF) – in which the particle is briefly at rest.[3] In this frame, where $\mathbf{v} = \mathbf{0}$, we have $\mathbf{U} = (1, \mathbf{0})$ and $\mathbf{A} = (\dot{\gamma}, \mathbf{a}) = (0, \mathbf{a})$ (since $\dot{\gamma}$ contains a factor of $\mathbf{v}$), so that

$$\mathbf{U} \cdot \mathbf{A} = 0$$

in this co-moving frame, and therefore in all frames. From the result in this co-moving frame we can deduce the magnitude of the 4-acceleration

$$\mathbf{A} \cdot \mathbf{A} = -a^2,$$

defining the *proper acceleration* $a$ as the magnitude of the acceleration in the instantaneously co-moving inertial frame. See also Exercise 6.11.

Finally, given two particles with velocities $\mathbf{U}$ and $\mathbf{V}$, and given that the second has velocity $v$ with respect to the first, then in the first particle's rest frame the velocity vectors have components $\mathbf{U} = (1, \mathbf{0})$ and $\mathbf{V} = \gamma(v)(1, \mathbf{v})$. Thus

$$\mathbf{U} \cdot \mathbf{V} = \gamma(v),$$

and this scalar product is, again, frame-independent. [Exercise 6.6]

---

[3] This is also referred to as the *momentarily co-moving reference frame* (MCRF) by some authors.

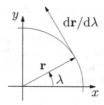

**Figure 6.4** The derivative of a vector is the tangent to its curve.

## 6.4 Velocities and Tangent Vectors

An alternative route to the velocity vector is to use the idea of a *tangent vector*. This is how vectors are defined in GR, and is a more fundamental approach to vectors than the one described above. This section is itself something of a tangent, which expands on the mention of 'tangent vectors' in the paragraph following Eq. (6.14). The account here is still rather compressed, and I offer it only to provide a hint of the more abstract, but more beautiful and powerful, way in which these things are handled in GR.

Consider a vector $\mathbf{r} = (x, y)$ on the euclidean plane. If the components $x$ and $y$ are functions of some parameter $\lambda$, then the vector function $\mathbf{r}(\lambda)$ will trace out a path $(x(\lambda), y(\lambda))$ on the plane. If we *differentiate* these components with respect to the parameter $\lambda$, then we will obtain an object which obviously tells us something about the path. For example, if we follow the path $\mathbf{r}(\lambda) = (\cos\lambda, \sin\lambda)$, parameterised by $\lambda$, we find it traces out a circle. Differentiating this straightforwardly, we find

$$\frac{d\mathbf{r}}{d\lambda} = (-\sin\lambda, \cos\lambda),$$

which is a vector which, when plotted at the position $\mathbf{r}(\lambda)$, as in Figure 6.4, can be clearly seen to be tangent to the path.

Now consider the spacetime vector $\mathbf{R} = (t(\lambda), x(\lambda), y(\lambda), z(\lambda))$, which draws out a path in spacetime. The path this traces out is the *worldline* – the set of events which take place along a moving particle's path through spacetime – and a reasonable parameter to use is the particle's proper time, $\tau$ – the time showing on the face of a clock attached to the moving particle. We therefore have a path $\mathbf{R} = (x^0(\tau), x^1(\tau), x^2(\tau), x^3(\tau))$ and, exactly as we did on the euclidean plane above, can differentiate it to obtain

$$\frac{d\mathbf{R}}{d\tau} = \left(\frac{dx^0}{d\tau}, \frac{dx^1}{d\tau}, \frac{dx^2}{d\tau}, \frac{dx^3}{d\tau}\right),$$

(compare Eq. (6.9)) as a 4-vector tangent to the worldline, which clearly

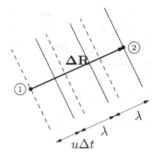

**Figure 6.5** A wavetrain: waves of wavelength $\lambda$ moving at speed $u$. In time $\Delta t$, the wavecrests move from their initial position (dashed) to an advanced one (solid).

contains information about the 'speed' of the particle, and which we can *define* to be the velocity 4-vector. By the same argument that led up to Eq. (6.14), we discover that this vector has magnitude $\mathbf{U} \cdot \mathbf{U} = 1$, and that its direction corresponds precisely to the $t$-axis of a frame co-moving with the particle.

The point of this approach is that the idea of a *path*, and the idea of a *tangent vector* to that path, are both *geometrical* ideas, existing at a level beneath coordinates (which are more-or-less algebraic things), and so can be defined and discussed without using coordinates, and so without having any dependence on reference frames. They are therefore manifestly frame-independent.

## 6.5 The Frequency Vector, and the Doppler Shift

In this section, we will examine a particular (non-obvious) 4-vector, and exploit its properties to deduce the relativistic Doppler effect.

Imagine a series of waves of some type (not necessarily light waves), moving in a direction $\mathbf{n} = (l, m, n)$ at a speed $u$, and imagine following a point on the crest of one of these waves; take the vector $\mathbf{n}$ to be a unit vector, so $\mathbf{n} \cdot \mathbf{n} = 1$. The point will have a displacement $(\Delta x, \Delta y, \Delta z)$ in time $\Delta t$, and so we will have

$$l\Delta x + m\Delta y + n\Delta z = u\Delta t,$$

for two events on the same wavecrest.

Now imagine a whole train of such waves, separated by wavelength $\lambda$, as illustrated in Figure 6.5. Consider the separation between one event on

the crest of a wave and another on a wavecrest $N$ (integer) wavecrests away
(that is, a different wavecrest), and write this as $\Delta R = (\Delta t, \Delta x, \Delta y, \Delta z)$. We
can observe that the time separation between events 1 and 2 is still $\Delta t$, so
that the wavetrain will have shifted a distance $u\Delta t$, meaning that the *spatial*
separation between the two events is $u\Delta t + N\lambda$. But this spatial separation
is also, as before, $l\Delta x + m\Delta y + n\Delta z$, so that we can write

$$l\Delta x + m\Delta y + n\Delta z = u\Delta t + N\lambda$$

or

$$u\Delta t - l\Delta x - m\Delta y - n\Delta z = -N\lambda. \tag{6.17}$$

Defining the frequency $f = u/\lambda$, we can rewrite this as

$$\mathbf{L} \cdot \Delta\mathbf{R} = -N, \tag{6.18}$$

where the dot represents the scalar product of Eq. (6.6), and the *frequency
4-vector* is

$$\mathbf{L} = \left(f, \frac{\mathbf{n}}{\lambda}\right) = \left(f, \frac{l}{\lambda}, \frac{m}{\lambda}, \frac{n}{\lambda}\right). \tag{6.19}$$

We have written this as a vector, but what is there to say that this is *really*
a vector – that is, that it is the components in one frame of an underlying
geometrical object – and isn't merely four numbers in a row? We know that
$\Delta\mathbf{R}$ is a vector – it is the prototype 4-vector – so we know that its components
transform according to Eq. (6.2b): $\Delta t = \gamma(\Delta t' + v\Delta x')$, $\Delta x = \gamma(\Delta x' + v\Delta t')$,
$\Delta y = \Delta y'$ and $\Delta z = \Delta z'$. We can substitute this into Eq. (6.18), rearrange to
gather terms $\Delta t'$, $\Delta x'$, $\Delta y'$, $\Delta z'$ and, just as we did before Eq. (6.18), rewrite
to find

$$\mathbf{L} \cdot \Delta\mathbf{R} = \left[\gamma\left(f - v\frac{l}{\lambda}\right), \gamma\left(\frac{l}{\lambda} - vf\right), \frac{m}{\lambda}, \frac{n}{\lambda}\right] \cdot \Delta\mathbf{R}'.$$

But the vector in the square brackets is exactly the vector $\mathbf{L}'$ we would obtain
if we transformed the frequency vector $\mathbf{L}$ according to the transformation
matrix Eq. (6.3). That is, we have spotted that $\mathbf{L}' = \Lambda\mathbf{L}$, and thus established
that

$$\mathbf{L}' \cdot \Delta\mathbf{R}' = \mathbf{L} \cdot \Delta\mathbf{R} = -N,$$

or in other words that it is a frame-invariant quantity. We have therefore
established that the object $\mathbf{L}$ defined by Eq. (6.19) really *is* a 4-vector, since it
transforms in the same manner as the prototype 4-vector $\Delta\mathbf{R}$ (there is no
significance to the value being negative: the important thing is that this is a
frame-invariant scalar).

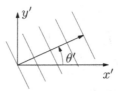

**Figure 6.6** A wavetrain: moving at an angle $\theta'$ to the $x'$-axis.

Is this not inevitable? Not quite: imagine if we had naïvely defined the frequency 4-vector as a vector whose space components were $\mathbf{n}/\lambda$ and whose time component was defined to be zero. On transformation by either of the routes in the previous paragraph, the vector would acquire a non-zero time component, so that the transformed vector would have a different *form* from the untransformed one. The components of such a 'vector' would *not* transform in the same way as $\Delta \mathbf{R}$, so it would not be a proper 4-vector, so that we would not be able to identify an underlying geometrical object of which these were the components.

Can we use the frequency 4-vector for anything? Yes. Imagine that the wavetrain is moving at an angle $\theta'$ in the $(x', y')$ plane, so that its direction is $\mathbf{n} = (\cos \theta', \sin \theta', 0)$ for some angle $\theta'$ in frame $S'$ (Figure 6.6). In that case we have

$$\mathbf{L'} = \left[ f', \frac{\cos \theta'}{\lambda'}, \frac{\sin \theta'}{\lambda'}, 0 \right] \tag{6.20}$$

$$\mathbf{L} = \left[ \gamma \left( f' + v \frac{\cos \theta'}{\lambda'} \right), \gamma \left( \frac{\cos \theta'}{\lambda'} + v f' \right), \frac{\sin \theta'}{\lambda'}, 0 \right]. \tag{6.21}$$

Now, by direct analogy with Eq. (6.20), write

$$\mathbf{L} = \left[ f, \frac{\cos \theta}{\lambda}, \frac{\sin \theta}{\lambda}, 0 \right], \tag{6.22}$$

and compare Eq. (6.21) and (6.22) component by component. After a bit of rearrangement, we find

$$f = f' \gamma \left( 1 + \frac{v}{u'} \cos \theta' \right) \tag{6.23}$$

$$\cos \theta = \frac{\cos \theta' + v u'}{1 + v u' \cos \theta'}. \tag{6.24}$$

Or, for the case of light, where $u' = 1$, we have the simpler and well-known versions

$$f = f'\gamma(1 + v\cos\theta') \tag{6.25}$$

$$\cos\theta = \frac{\cos\theta' + v}{1 + v\cos\theta'}. \tag{6.26}$$

Equation (6.23) is the *relativistic Doppler effect*, and describes the change in frequency of a wave, as measured in a frame moving with respect to the frame in which it was emitted. This applies for everything from water waves (for which the effect would be exceedingly small) all the way up to light, for which $u = 1$.

Equation (6.24) shows that a wave travelling at an angle $\theta'$ in the moving frame $S'$ is measured to be moving at a *different* angle $\theta$ in a frame $S$, with respect to which $S'$ is moving with speed $v$. To calculate the change in the speed of the wave, we could laboriously eliminate variables from Eq. (6.21) and (6.22), but much more directly, we can make use of the fact that the magnitudes of vectors are conserved under Lorentz transformation; thus $\mathbf{L} \cdot \mathbf{L} = \mathbf{L}' \cdot \mathbf{L}'$ or, again using Eq. (6.5) and $f = u/\lambda$,

$$f^2\left(1 - \frac{1}{u^2}\right) = f'^2\left(1 - \frac{1}{u'^2}\right).$$

We could rewrite this to obtain an expression for $u'$, but simply from this form we can see that if $u = 1$, the fact that neither $f$ nor $f'$ is zero implies that $u' = 1$ also (as the second postulate says).

Note that there are multiple possible ways to set up this problem, from a notational point of view, so that if you are comparing the discussion here with another text, make sure the various symbols mean what you expect.

[Exercises 6.7–6.17]

## Exercises

### Exercise 6.1 (§6.2)

Verify that the transformation matrices of Eq. (6.3a) and Eq. (6.3b) are inverses. You will probably find it convenient to omit the (trivial) $y$ and $z$ components, and instead write

$$\Lambda = \gamma\begin{pmatrix} 1 & -v \\ -v & 1 \end{pmatrix}.$$

Observe additionally that $\Lambda$ is obtained from $\Lambda^{-1}$ by negating $v$.          [ $d^-$ ]

## Exercise 6.2 (§6.2)

Verify Eq. (6.8). As in Exercise 6.1, ignore the trivial components and write

$$\eta = \begin{pmatrix} 1 & 0 \\ 0 & -1 \end{pmatrix}.$$

You can write this as a matrix expression, or as a sum over components.

$[d^-]$

## Exercise 6.3 (§6.2)

Consider the vectors:

- $\mathbf{A} = (1, 2, 0, 0)$
- $\mathbf{B} = (2, 1, 0, 0)$

Calculate $\mathbf{A} \cdot \mathbf{B}$, and the magnitudes of $\mathbf{A}$ and $\mathbf{B}$. $[d^-]$

## Exercise 6.4 (§6.2)

Consider the vectors $\mathbf{A} = (5, 3, 4, 0)$, $\mathbf{B} = (1, 1, -1, -1)$ and $\mathbf{C} = (3, 2, -1, 0)$. Calculate $\mathbf{A} \cdot \mathbf{B}$, $\mathbf{A} \cdot \mathbf{C}$ and $\mathbf{B} \cdot \mathbf{C}$: which pair is orthogonal? Calculate also the magnitudes of $\mathbf{A}$, $\mathbf{B}$ and $\mathbf{C}$, and indicate which vectors are null, timelike and spacelike.

## Exercise 6.5 (§6.2.1)

Consider, instead of Eq. (6.2a), the transformation

$$\begin{pmatrix} \Delta t' \\ \Delta x' \\ \Delta y' \\ \Delta z' \end{pmatrix} = \begin{pmatrix} a & b & 0 & 0 \\ c & d & 0 & 0 \\ 0 & 0 & 1 & 0 \\ 0 & 0 & 0 & 1 \end{pmatrix} \begin{pmatrix} \Delta t \\ \Delta x \\ \Delta y \\ \Delta z \end{pmatrix} \qquad \text{(i)}$$

which is the simplest transformation which 'mixes' the $x$- and $t$-coordinates. By *requiring* that $\Delta s'^2 = \Delta s^2 = \Delta t^2 - \Delta x^2$ after this transformation, find constraints on the parameters $a, b, c, d$ (you can freely add the constraint $b = c$; why?), and by setting $b$ proportional to $a$ deduce the matrix Eq. (6.2a).

Bonus: by instead parameterising $a, b, c, d$ with suitable hyperbolic functions, recover Eq. (5.3). $[d^+]$

## Exercise 6.6 (§6.3)

I throw a ball vertically into the air. At the top of its path, what is the instantaneously comoving reference frame?

- The frame of the room.
- A frame moving upwards with the same speed as I threw the ball.
- A frame moving downwards with the ball's terminal velocity.      [ $d^-$ ]

## Exercise 6.7 (§6.5)

Consider a water wave moving along the $x$-axis at speed $10\,\mathrm{m\,s^{-1}}$, with wavelength $5\,\mathrm{m}$. What is its frequency 4-vector?

- $(2, 2, 0, 0)$
- $(2, 0.2, 0, 0)$
- $(10, 5, 0, 0)$                                                   [ $d^-$ ]

## Exercise 6.8 (§6.5)

One might naïvely consider defining a 4-vector representing the electric field as $(0, E^x, E^y, E^z)$, where the $E^i$ are the components of the electric field, and the time component is defined to be identically zero. Explain briefly why this *cannot* be a 4-vector.      [ $u^{++}$ ]

## Exercise 6.9 (§6.5)

Obtain the inverse to Eq. (6.25) and Eq. (6.26) by applying Eq. (6.3) to Eq. (6.22) and comparing with Eq. (6.20). Verify that the resulting expression matches that which you obtain through the procedure described in Section 6.2: swapping primed with unprimed quantities and changing the sign of $v$.

## Exercise 6.10 (§6.5)

Obtain Eq. (6.24) by observing that $\tan\theta = L^2/L^1$ (the $x$ and $y$ components of $\mathbf{L}$) and doing the trigonometry. You should recall that $\sec^2 x = 1 + \tan^2 x$, and it will be useful to notice that $1 = \gamma^2(1 - v^2)$.

## Exercise 6.11 (§6.5)

A rocket starts at rest at a space station, and accelerates along the $x$-axis at a constant rate $\alpha$ in its own frame $S'$ (i.e., in the ICRF). The rocket's (proper) acceleration 4-vector is therefore $A' = (0, \alpha, 0, 0)$.

(a) Using the (inverse) Lorentz transformation equations, to transform from an inertial frame instantaneously co-moving with the rocket into the space station frame, calculate the components of the rocket's acceleration vector in the space station's frame, and confirm that $\mathbf{A} \cdot \mathbf{A} = -\alpha^2$ in this frame also. Write down the rocket's velocity 4-vector in this frame, and confirm that $\mathbf{U} \cdot \mathbf{A} = 0$. Differentiate this with respect to the proper time, to obtain

$$\frac{d\mathbf{U}}{d\tau} = \gamma \left( \frac{d\gamma}{dt}, \frac{d}{dt}(\gamma v), 0, 0 \right).$$

Hence deduce that $\gamma v = \alpha t$, and thus that

$$\frac{1}{v^2} = \frac{1}{\alpha^2 t^2} + 1. \tag{i}$$

(b) If, at time $t$, the rocket sets off a flashbulb which has frequency $f'$ in its frame, use the Doppler formula to show that the light is observed, at the space station, to have frequency

$$f = \left( \sqrt{1 + \alpha^2 t^2} - \alpha t \right) f'.$$

What is this factor if the flashbulb is set off at time $t = 3/(4\alpha)$?

(c) Rearrange Eq. (i) to produce an expression for $v$, write $v = dx/dt$, and integrate to obtain the constant-acceleration equation for SR,

$$\left( x + \frac{1}{\alpha} \right)^2 - t^2 = \frac{1}{\alpha^2}$$

(you may or may not find the substitution $\alpha t = \sinh \theta$ useful or obvious here). This is the equation for a hyperbola. By sketching this on a Minkowski diagram and considering the point at which the asymptote intersects the $t$-axis, demonstrate that it is impossible for the space station to signal to the retreating rocket after a time $1/\alpha$.

(d) How long does it take, from launch at $v = 0$, for a rocket accelerating at $g = 10 \, \mathrm{m\,s^{-2}}$ to be moving at $\gamma = 2$?

[This question is a little algebra-heavy, but instructive.]          $[\,d^+ u^+\,]$

### Exercise 6.12 (§6.5)

Consider an ambulance passing you with its siren sounding. At the instant the ambulance is level with you, its siren has its 'rest' frequency.

If the ambulance were moving at a relativistic speed, what would the siren sound like as it passed you? Would it be:

- different from its rest frequency?
- the same as its rest frequency? $[d^-]$

### Exercise 6.13 (§6.5)

Consider (again) an ambulance passing you with its blue light flashing.

If the ambulance were moving at a relativistic speed, what would be the measured frequency of the ambulance's blue light, as a function of its frequency in its rest frame, $f'$? If we put $\theta' = \pi/2$ into Eq. (6.25) (examining the light which is emitted perpendicular to the ambulance's motion), then we obtain $f = \gamma f'$. But this suggests that the frequency in our frame is measured to be *higher* than the frequency in the rest frame, and appears to directly contradict the calculations in Exercise 5.2. What is wrong with this argument? $[d^+]$

### Exercise 6.14 (§6.5)

*The traditional traffic-lights example*: You are driving towards some traffic lights showing red (take wavelength $\lambda = 600$ nm). How fast do you have to be driving so that they are Doppler-shifted enough that they appear green (wavelength $\lambda = 500$ nm)? [Hint: rewrite Eq. (6.23) in terms of $\rho = f'/f$, set $u = 1$ and $\theta = 0$, and rearrange to make $v$ the subject.] Does this sound like a reasonable defence when you're had up in court on a charge of dangerous driving?

### Exercise 6.15 (§6.5)

Light emitted from a distant star may be Doppler-shifted, so that we detect it at a frequency which is different from the frequency at which it was emitted (for example, the star may be a member of a binary system). You may have come across this Doppler shift expressed as

$$f = \left(1 + \frac{v}{c}\right) f'.$$

This is not the same as Eq. (6.25), even for $\theta' = 0$. Why not? Derive this expression from Eq. (6.25).

### Exercise 6.16 (§6.5)

In 1725, James Bradley measured an aberration in the altitude of a star that was dependent on the Earth's motion at a given point in its orbit. The observational result of Bradley's investigation is

$$\Delta\theta = \kappa \sin\theta$$

where $\Delta\theta$ is the aberration in the star's altitude $\theta$, and $\kappa = 20''.496$ is the constant of aberration.

Consider pointing a laser at a star which has altitude $\theta$ in a frame in which the Earth is at rest, and altitude $\theta'$ in a frame which is moving at the Earth's orbital velocity of approximately $v = 29.8$ km s$^{-1}$ (the two frames are equivalent to making an observation where the Earth's motion is orthogonal to the line-of-sight to the star, and a frame, three months later, which is moving towards the star; also, it's easier to solve this variant of the problem rather than the equivalent problem of the light coming from the star, since it avoids lots of minus signs in the algebra). Show that the two altitudes are related by:

$$\cos\theta = \frac{\cos\theta' + v}{1 + v\cos\theta'}. \tag{i}$$

Given this, it is merely tedious to work out that $\sin\theta = \sin\theta'/(\gamma(1 + v\cos\theta'))$.

Deduce that, for small $v$,

$$\Delta\theta = \theta' - \theta \approx v\sin\theta',$$

and verify that this is consistent with Bradley's observational result.

The main challenge to this question is keeping track of the trigonometry, but it's useful drill nonetheless. [Recall that $\tan^2\theta = 1/\cos^2\theta - 1$, that $\alpha \approx \sin\alpha$ for small $\alpha$, and that $\sin(\alpha - \beta) = \sin\alpha\cos\beta - \cos\alpha\sin\beta$.]

$$[d^+]$$

### Exercise 6.17 (§6.5)

Recall the expression for $\cos\theta$ in Exercise 6.16.

A source in the moving frame $S'$ emits light isotropically. Deduce that light emitted into the forward half-sphere (that is, with half-angle $\theta' \leq \theta'_+ =$

$\pi/2$) is observed, in the frame $S$, to be emitted into a cone with half-angle $\theta_+$, where $\cos\theta_+ = v$. This is known as the *headlight effect*.

# 7

# Dynamics

In the previous chapter, we have learned how to describe motion; we now want to explain it. In newtonian mechanics, we do this by defining quantities such as momentum, energy, force and so on. To what extent can we do this in the context of relativity, with our new 4-vector tools?

*Aims*: you should:

7.1. understand relativistic energy and momentum, the concept of energy-momentum as the magnitude of the momentum 4-vector, and conservation of the momentum 4-vector; and
7.2. understand the distinction between invariant, conserved and constant quantities.

## 7.1 Energy and Momentum

We can start with momentum. We know that in newtonian mechanics, momentum is defined as mass times velocity. We have a velocity, so we can try defining a momentum 4-vector, for a particle of mass $m$ moving at velocity $\mathbf{v}$, as

$$\mathbf{P} = m\mathbf{U} = m\gamma(1, \mathbf{v}). \tag{7.1}$$

Since $m$ is a scalar, and $\mathbf{U}$ is a 4-vector, $\mathbf{P}$ must be a 4-vector also. Remember also that $\gamma$ is a function of the 3-vector $\mathbf{v}$: $\gamma(\mathbf{v})$.

In the rest frame of the particle, this becomes $\mathbf{P} = m(1, \mathbf{0})$: it is a 4-vector whose length ($\sqrt{\mathbf{P} \cdot \mathbf{P}}$) is $m$, and which points along the particle's worldline. That is, it points in the direction of the particle's movement in spacetime. Since this is a vector, its magnitude and its direction are frame-independent

125

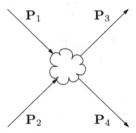

**Figure 7.1** The momenta involved in a collision.

quantities, so a particle's 4-momentum vector *always points in the direction of the particle's worldline*, and the 4-momentum vector's *length is always m*. We'll call this vector the *momentum (4-)vector*, but it's also called the *4-momentum* or the *energy-momentum vector*, and Taylor & Wheeler (coining the word in an excellent chapter on it) call it the *momenergy vector* in order to stress that it is *not* the same thing as the energy or momentum (or mass) that you are used to.

> Note that here, and throughout, the symbol $m$ denotes the mass as measured in a particle's rest frame. The reason I mention this is that some treatments of relativity, particularly older ones, introduce the concept of the 'relativistic mass' $m(v) = \gamma(v)m_0$, distinct from the 'rest mass', $m_0$. The only (dubious) benefit of this is that it makes a factor of $\gamma$ disappear from a few equations, making them look a little more like their newtonian counterparts; the cost is that of introducing one more new concept to worry about, which doesn't help much in the long term, and which can obscure aspects of the energy-momentum vector. Rindler (2006, §6.2), for example, introduces the relativistic mass, but his subsequent discussion of relativistic force is sufficiently confusing, from our notational point of view (it obliges one to introduce the notions of 'longitudinal' and 'transverse relativistic mass'!) that I feel it quite amply illustrates the unhelpfulness of the whole idea.

> The discussion of dynamics, in this chapter, also draws on chapter 2 of Schutz (2009). This is a textbook on General Relativity, but chapter 2 describes Special Relativity using the beautiful and powerful geometrical language which is used extensively in GR. I have toned this down for this section, but the motivation is the same. It also draws on the insightful discussion in Taylor & Wheeler (1992). 'Insightful' doesn't really do justice to Taylor & Wheeler – you should read chapter 7 of the book if you possibly can.

Now consider a pair of incoming particles $\mathbf{P}_1$ and $\mathbf{P}_2$ which collide and produce a set of outgoing particles $\mathbf{P}_3$ and $\mathbf{P}_4$ (see Figure 7.1; this can be trivially extended to more than two particles). Suppose that the total momentum

is conserved:

$$\mathbf{P}_1 + \mathbf{P}_2 = \mathbf{P}_3 + \mathbf{P}_4. \tag{7.2a}$$

This is an equation between 4-vectors. Equating the time and space coordinates separately, recalling Eq. (7.1), and writing $\mathbf{p} \equiv \gamma m \mathbf{v}$, we have

$$m_1 \gamma(v_1) + m_2 \gamma(v_2) = m_3 \gamma(v_3) + m_4 \gamma(v_4) \tag{7.2b}$$

$$\mathbf{p}_1 + \mathbf{p}_2 = \mathbf{p}_3 + \mathbf{p}_4. \tag{7.2c}$$

Now recall that, as $v \to 0$, we have $\gamma(v) \to 1$, so that the low-speed limit of the spatial part of the vector $\mathbf{P}$, Eq. (7.1), is just $m\mathbf{v}$, so that the spatial part of the conservation equation, Eq. (7.2c), reduces to the statement that $m\mathbf{v}$ is conserved. Both of these prompt us to identify the spatial part of the vector $\mathbf{P}$ as the linear momentum, and to retrospectively justify both giving the 4-vector $\mathbf{P}$ the name 4-momentum and supposing that it is conserved.

What, then, of the time component of Eq. (7.1)? Let us (with, admittedly, a little fore-knowledge) write this as $P^0 = E$, so that

$$E = \gamma m. \tag{7.3}$$

What is the low-speed limit of this? Taylor's theorem tells us that

$$\gamma = \left(1 - v^2\right)^{-1/2} = 1 + \frac{v^2}{2} + O(v^4), \tag{7.4}$$

so that, when $v$ is small, Eq. (7.3) becomes

$$E = m + \frac{1}{2} m v^2 + O(v^4). \tag{7.5}$$

At this point we can (a) spot that $m v^2 / 2$ is the expression for the kinetic energy in newtonian mechanics, and (b) recall that Eq. (7.2b) tells us that this quantity $E$ is conserved in collisions, so that we have persuasive support for identifying the quantity $E$ in Eq. (7.3) as the relativistic *energy* of a particle with mass $m$ and velocity $v$.

If we rewrite Eq. (7.3) in physical units, we find

$$E = \gamma m c^2, \tag{7.6}$$

the low-speed limit of which (remember $\gamma(0) = 1$) recovers what has been called the most famous equation of the twentieth century.

Note that the units in Eq. (7.3) are kg on both sides, but after rewriting in units where $c \neq 1$, Eq. (7.6), the units on both sides are $\mathrm{kg\,m^2\,s^{-2}}$, or joules, as expected.

The argument presented after Eq. (7.2a) has been concerned with giving names to quantities, and, reassuringly for us, linking those newly named things with quantities we already know about from newtonian mechanics. This may seem artificial, and it is certainly not any sort of proof that the 4-momentum is conserved as Eq. (7.2a) says it might be. What we are doing here, however, is that we are *postulating* that 4-momentum is conserved as in Eq. (7.2a), which allows us to deduce that energy and 3-momentum are separately conserved, as we expect. No proof is necessary, however: it turns out *from experiment* that Eq. (7.2a) is indeed a law of nature, and nothing more really needs to be said.

> In case you are worried that we are pulling some sort of fast one, that we never had to do in newtonian mechanics, note that we *do* have to do a similar thing in newtonian mechanics. There, we postulate Newton's third law (action equals reaction), and from this we can deduce the conservation of momentum; in this case, we work in the opposite direction, so that we postulate the conservation of 4-momentum, and would then be able to deduce a relativistic analogy of Newton's third law. In each case, we are adding the same amount of physics to the mathematics. I don't discuss relativistic force here, but merely mention it in passing in Section 7.6; this question is discussed in a little more detail in Rindler (2006, §6.10).

We can see from Eq. (7.5) that, even when a particle is stationary and $v = 0$, the energy $E$ is non-zero. In other words, a particle of mass $m$ has an energy $\gamma m$ associated with it simply by virtue of its mass.

The low-speed limit of Eq. (7.2b) simply expresses the conservation of mass, but we see from Eq. (7.3) that it is actually expressing the conservation of energy. In SR there is no fundamental distinction between mass and energy – mass is, like kinetic, thermal and strain energy, merely another form into which energy can be transmuted – albeit a particularly dense store of energy, as can be seen by calculating the energy equivalent, in joules, of a mass of 1 kg. It turns out from GR that it is not mass that gravitates, but energy-momentum (most typically, however, in the particularly dense form of mass), so that thermal and electromagnetic energy, for example, and even the energy in the gravitational field itself, all gravitate (it is the non-linearity implicit in the last remark that is part of the explanation for the mathematical difficulty of GR). In each of these cases, the amount of (thermal, electromagnetic, strain) energy would be a quantity with the units of kilogrammes.

Although the quantities $\mathbf{p} = \gamma m\mathbf{v}$ and $E$ are frame-dependent, and thus not physically meaningful by themselves, the quantity $\mathbf{P}$ defined by Eq. (7.1) has a physical significance.

Let us now consider the magnitude of the 4-momentum vector. Like any such magnitude, it will be frame-invariant, and so will express something fundamental about the vector, analogous to its length. Since this is the momentum vector we are talking about, this magnitude will be some important invariant of the motion, indicating something like the 'quantity of motion'. From the definition of the momentum, Eq. (7.1), and its magnitude, Eq. (6.14), we have

$$\mathbf{P} \cdot \mathbf{P} = m^2 \mathbf{U} \cdot \mathbf{U} = m^2, \tag{7.7}$$

and we find that this important invariant is the *mass* of the moving particle.

Now using the definition of energy, Eq. (7.3), we can write $\mathbf{P} = (E, \mathbf{p})$, and find

$$\mathbf{P} \cdot \mathbf{P} = E^2 - \mathbf{p} \cdot \mathbf{p}. \tag{7.8}$$

Writing now $p^2 = \mathbf{p} \cdot \mathbf{p}$, we can combine these to find

$$m^2 = E^2 - p^2. \tag{7.9}$$

Equation (7.9) appears to be simply a handy way of relating $E$, $p$ and $m$, but that conceals the more fundamental meaning of Eq. (7.7). The 4-momentum $\mathbf{P}$ encapsulates the important features of the motion, through the energy and spatial momentum. The quantities $E$ and $p$ are frame-dependent separately, but combine into a frame-independent quantity.

It seems odd to think of a particle's momentum as being always non-zero, irrespective of how rapidly it's moving; this is the same oddness that has the particle's velocity always being of length 1. One way of thinking about this is that it shows that the 4-momemtum vector (or energy-momentum or momenergy vector) is a 'better' thing than the ordinary 3-momentum: it's frame-independent, and so has a better claim to being something intrinsic to the particle. Another way of thinking about it is to conceive of the 4-velocity as showing the particle's movement through spacetime. In a frame in which the particle is 'at rest', the particle is in fact moving at speed $c$ into the future. If you are looking at this particle from another frame, you'll see the particle move in space (it has non-zero space components to its 4-velocity in your frame), and it will consequently (in order that $\mathbf{U} \cdot \mathbf{U} = 1$) have a *larger* time component in your frame than it has in its rest frame. In other words, the particle moves through more time in your frame than it does in its rest frame – yet another manifestation of time dilation. Multiply this velocity by the particle's mass, and you can imagine the particle moving into the future with a certain momentum in its rest frame; observe this particle from a moving frame and its spatial momentum becomes non-zero, and the time component of its momentum (its energy) has to become bigger – the particle packs more punch – as a consequence of the length of the momentum vector being invariant.

[Exercises 7.1–7.3]

### 7.1.1 Another Route to the Momentum 4-vector ◬

Above, we saw how we can define the momentum 4-vector in a rather obvious way, by defining it as $m\mathbf{U}$, in direct analogy with the newtonian momentum $m\mathbf{v}$. Another route to this, which brings out different points of note, is to work by analogy with the way we developed the frequency vector in Section 6.5, and ask 'what would a momentum 4-vector *have* to look like?'

Start by thinking of a mass $m$ at rest. What is its 4-momentum $\mathbf{Q} = (Q^0, \mathbf{q})$? Based on our experience in Chapter 6, we can suppose that the spatial part of a 4-momentum will be simply related to the particle's newtonian momentum, and in particular that it will reduce to the newtonian value $m\mathbf{v}$ at low speed, so that if the mass is at rest, $\mathbf{q} = 0$.

The obvious next step is to suppose that $Q^0 = 0$ for the particle at rest, but if a 4-vector has all-zero components in one frame, then Eq. (6.3a) tells us that it would have all-zero components in every frame, so this won't work. So we'll presume that $Q^0 \neq 0$, restrict ourselves to motion along the $x$-axis, and write

$$\mathbf{Q} = (Q^0, 0).$$

What does this 4-vector look like in another frame? In particular, let's transform it into the frame $S'$, which is moving with respect to the particle at speed $-v$ along the $x$-axis, so that the particle is moving at speed $+v$ in $S'$. Looking again at Eq. (6.3a), we find

$$Q'^0 = \gamma Q^0$$
$$Q'^1 = \gamma Q^0 v.$$

The low-speed limit of the second equation is (from Eq. (7.4)) $Q'^1 = Q^0 v + O(v^3)$. If this is to reduce to the newtonian momentum $Q'^1 = mv$ at low speed, then we must identify $Q^0 = m$. Looking now at the first expression, we have

$$Q'^0 = Q^0\left(1 + \frac{v^2}{2} + O(v^4)\right) = m + \frac{1}{2}mv^2 + O(v^4),$$

and spotting $mv^2/2$ again prompts us to interpret the zeroth component of the 4-momentum as an energy (or at least something closely related to newtonian energy), and recovers what we now call the (relativistic) energy of the moving particle as $Q'^0 = \gamma m$, as in Eq. (7.3).

This 4-momentum $\mathbf{Q}$ therefore has a plausible dynamical interpretation, to which we can add the supposition that the 4-momentum is conserved in collisions (a physical, rather than a mathematical statment), and pick up the argument after Eq. (7.6).

We have not added anything in this section that we didn't know from the previous one. But simply by demanding that a quantity of interest is a 4-vector, as we did here and in Section 6.5, we have imposed a significant structure on that quantity, and discovered that two quantities that newtonian physics sees as related but distinct – frequency and wavelength, or energy and momentum – are in SR different components of a single geometrical quantity. We did the same thing in Section 6.3 when we remarked that we could define the 4-velocity of a particle by simply declaring it to be the vector which had components $(1, \mathbf{0})$ in the particle's rest frame, and letting Eq. (6.3a) do the rest of the work. This shows the fruitfulness of an approach which focuses on the *geometry* of the spacetime we are examining; such an approach may feel a little abstract, but it pays off when we look at the description of gravity in general relativity.

## 7.2 Photons

For a photon, the interval represented by $d\mathbf{R} \cdot d\mathbf{R}$ is always zero ($d\mathbf{R} \cdot d\mathbf{R} = dt^2 - dx^2 - dy^2 - dz^2 = 0$ for photons). But this means that the proper time $d\tau^2$ is also zero for photons. This means, in turn, that we cannot define a 4-velocity vector for a photon by the same route that led us to Eq. (6.9), and therefore cannot define a 4-momentum in the way we did in Eq. (7.1).

We can do so, however, by a different route. Recall that we defined (in the paragraph below Eq. (6.14)) the 4-velocity as a vector pointing along the worldline, which resulted in the 4-momentum being in the same direction. From the discussion of the momentum of massive particles above, we see that the $P^0$ component is related to the energy, so we can use this to define a 4-momentum for a *massless* particle, and again write

$$\mathbf{P}_\gamma = (E, \mathbf{p}_\gamma).$$

Since the photon's velocity 4-vector is null, the photon's 4-momentum must be also (since it is defined above to be pointing in the same direction). Thus we must have $\mathbf{P}_\gamma \cdot \mathbf{P}_\gamma = 0$, thus $\mathbf{p}_\gamma \cdot \mathbf{p}_\gamma = E^2$, recovering the $m = 0$ version of Eq. (7.9),

$$E^2 = p^2 \qquad \text{(massless particle)}, \qquad (7.10)$$

so that even massless particles have a non-zero momentum.

In quantum mechanics, we learn that the energy associated with a quantum of light – a photon – is $E = hf$, where $h$ is Planck's constant,

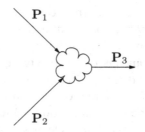

**Figure 7.2** The momenta involved in a collision.

$h = 6.626 \times 10^{-34}$ J s (or $2.199 \times 10^{-42}$ kg m in natural units), so that

$$\mathbf{P} = (hf, hf, 0, 0) \qquad \text{(photon)}. \tag{7.11}$$

Thinking back to the approach of Section 7.1.1, we realise that Eq. (7.11) is the conclusion we must come to if we speculate that the zeroth component of a photon's 4-momentum is its energy, $E = hf$, and then demand that that 4-vector is null.

## 7.3 Relativistic Collisions and the Centre-of-Momentum Frame

Consider two particles, of momenta $\mathbf{P}_1$ and $\mathbf{P}_2$, which collide and produce a single particle of momentum $\mathbf{P}_3$ – you can think of this as describing either two balls of relativistic putty, or an elementary particle collision which produces a single new outgoing particle (this example is adapted from the characteristically *excellent* discussion in Taylor & Wheeler (1992, §8.3)); note also that the subscripts denote the different particle momenta, but the superscripts denote the different momentum components, so that (for example) $P_2^1$ is component 1 (the 'x-component') of vector $\mathbf{P}_2$.

Recalling Eq. (7.1), the particles have momenta

$$\mathbf{P}_i = \gamma_i m_i (1, \mathbf{v}_i), \tag{7.12}$$

where the three particles have velocities $\mathbf{v}_i$, and $\gamma_i \equiv \gamma(v_i)$. From momentum conservation, we also know that

$$\mathbf{P}_1 + \mathbf{P}_2 = \mathbf{P}_3. \tag{7.13}$$

Table 7.1 *The momenta of the particles in Figure 7.2, in the 'lab frame'*

|   | $\mathbf{P}_i = (E_i, p_i)$ | $v_i$ | $\gamma_i$ | $m_i$ |
|---|---|---|---|---|
| 1 | $(17, 15)$ | $15/17$ | $17/8$ | $\sqrt{17^2 - 15^2} = 8$ |
| 2 | $(8, 0)$ | $0$ | $1$ | $8$ |
| 3 | $(17 + 8, 15 + 0)$ $= (25, 15)$ | $3/5$ | $5/4$ | $\sqrt{25^2 - 15^2} = 20$ |

Of course, this also indicates that *each component* of the vectors is separately conserved.

It's useful to add some numbers to the discussion at this point.

Let's simplify this, and imagine the collision taking place in one dimension, with an incoming particle moving along the $x$-axis to strike a stationary second particle. The resulting particle will also move along the $x$-axis. In this subsection we will, to avoid clutter, write only the $x$-component of the spatial part of vectors, missing out the $y$ and $z$ components, which are zero in this one-dimensional setup. Thus we'll write $(t, x)$ rather than $(t, x, y, z)$ or $(t, \mathbf{r})$, and $(P^0, P^1)$ rather than $(P^0, P^1, P^2, P^3)$.

Let the 'incoming' particles have masses $m_1 = m_2 = 8$ units; let the first be travelling with speed $v_1 = 15/17$ along the $x$-axis in the 'lab frame' (so that $\gamma_1 = \gamma(15/17) = 17/8$); and let the second be stationary, $v_2 = 0$ (so $\gamma_2 = 1$). As discussed above, the appropriate measure of the 'length' of the $\mathbf{P}$ vectors is the magnitude $m^2$ of Eq. (7.7): $m^2 = \mathbf{P} \cdot \mathbf{P}$. Recall that $\mathbf{P}_i = (E_i, p_i) = (\gamma_i m_i, \gamma_i m_i v_i)$. Thus we can make a table of the various kinematical parameters in Table 7.1, where the first two rows are simply transcribed from this paragraph.

To obtain the values in row 3, we add the $\mathbf{P}_i$ of the first column to get a $\mathbf{P}_3 = (25, 15)$. We can get the speed by noticing that $v_i = P_i^1/P_i^0$, from Eq. (7.12), and get the mass via the magnitude of the momentum 4-vector, $m^2 = (P^0)^2 - (P^1)^2$ (that's the most straightforward route; alternatively we could note that if $v_3 = 3/5$, then $\gamma_3 = 5/4$, so that if $P_3^0 = \gamma_3 m_3 = 25$, then $m_3 = 20$).

Notice the following points.

1. The two incoming particles have energy-momentum vectors with the same magnitude, $m_1 = m_2 = 8$. This may seem surprising, since one is moving much faster than the other, but recall that the scalar product $\mathbf{P}_1 \cdot$

$\mathbf{P}_1$ is an *invariant*, so it has the same value in the frame in which particle 1 is moving at speed $v_1 = 15/17$ as it has in the frame in which particle 1 is stationary; in the latter frame, $\mathbf{P}_1' = (\gamma(0)m_1, v_1) = (8, 0)$, showing the same values as $\mathbf{P}_2$ in the lab frame.

2. The mass in the 'outgoing' particle 3 is *not* merely the sum of the two incoming masses: $m_1 + m_2 \neq m_3$. It is the components of $\mathbf{P}$, including total energy $P_1^0 + P_2^0 = P_3^0 = \gamma m_3$, that are the quantities conserved in collisions.

3. The quantity $m^2 = \mathbf{P} \cdot \mathbf{P}$ is frame-invariant, but not conserved; $P^\mu$ is conserved, but not frame-invariant.

We must carefully distinguish the terms 'conserved', 'covariant', 'invariant' and 'constant' (cf. Taylor & Wheeler (1992, Box 7-3)):

**Conserved** refers to a quantity which is not changed by some process (most typically unchanged in time) – momentum, for example, is conserved in a collision. This refers to one frame at a time, and a conserved quantity will typically have different numerical values in different frames.

**Covariant** refers to an equation which does not change its *form* when it is transformed from one frame to another.

**Invariant** refers to a quantity which is not changed by some transformation – the radius $r$ is an invariant of the rotation in Figure 4.9, and we have just discovered that $m^2$ is an invariant of the Lorentz transformation. The term necessarily refers to more than one reference frame; there is no reason to expect that an invariant quantity will be conserved in a particular process. We will also sometimes refer to this as 'frame-independent'.

**Constant** refers to a quantity which does not change in time, such as the mass of a test particle.

The speed of light is one of the very few things which has the full-house of being conserved, invariant, *and* constant.                    [Exercises 7.4 & 7.5]

### 7.3.1 The Centre-of-Momentum Frame

From Eq. (7.12) and Eq. (7.13) we can write

$$\mathbf{P}_3 = (\gamma_1 m_1 + \gamma_2 m_2, \gamma_1 m_1 v_1 + \gamma_2 m_2 v_2).$$

Table 7.2 *The momenta of the particles in Figure 7.2, in the 'centre-of-momentum frame'*

| | $\mathbf{P}'_i = (E'_i, p'_i)$ | $v_i$ | $m_i$ |
|---|---|---|---|
| 1 | $(10, 6)$ | $3/5$ | $\sqrt{10^2 - 6^2} = 8$ |
| 2 | $(10, -6)$ | $-3/5$ | $8$ |
| 3 | $(20, 0)$ | $0$ | $20$ |

Now use the LT, Eq. (6.3), to transform this to a frame which is moving with speed $V$ with respect to the lab frame. We therefore have

$$P'^1_3 = \gamma(V)(P^1_3 - VP^0_3)$$
$$= \gamma(V)(\gamma_1 m_1 v_1 + \gamma_2 m_2 v_2 - V(\gamma_1 m_1 + \gamma_2 m_2)).$$

We can choose the speed $V$ to be such that this spatial momentum is zero, $P'^1_3 = 0$, giving

$$V = \frac{\gamma_1 v_1 + \gamma_2 v_2}{\gamma_1 + \gamma_2} = \frac{3}{5}$$

(since here $m_1 = m_2$). We want to obtain $\mathbf{P}'_i$ by starting with the energy momentum vectors $\mathbf{P}_i$ in the table above, and Lorentz-transforming them into this new frame. Using $\gamma(V = 3/5) = 5/4$, we have for example $P'^0_i = \gamma(P^0_i - VP^1_i))$ in Table 7.2.

Notice:

1. In this frame also, $\mathbf{P}'_1 + \mathbf{P}'_2 = \mathbf{P}'_3$: the conservation of a vector quantity in one frame means that it is conserved in all frames.
2. Also $m'_1 = (\mathbf{P}'_1 \cdot \mathbf{P}'_1)^{1/2} = 8$, $m'_2 = 8$ and $m'_3 = 20$: these are frame-invariant quantities.
3. Both the 3-momenta *and* energy of the vectors in this frame are different from the lab frame, even though their magnitudes are the same. We are talking about *the same 4-vector* here, $\mathbf{P}_i$, but since we have decided to change reference frame, we (unsurprisingly) change the vector's frame-dependent components.

This frame, in which the total spatial momentum is zero, so that the incoming particles have equal and opposite spatial momenta, is known as the *centre-of-momentum* (CM) frame, and the energy available for particle production in this frame ($P^0_{CM}$) is known as the *centre-of-mass energy*.

[Exercise 7.6]

## 7.4 But Where's This Mass Coming From?

In the collision discussed in Section 7.3.1, we saw $m_1 + m_2 \neq m_3$. We explained this by noting that although the rest masses of the before and after particles are frame-invariant, they are not conserved in the collision.

What's going on here? How much mass is actually present here? Put in physical terms, is this telling us that there would be more *gravity* after the collision than before? Excellent questions.

Imagine putting a box round the collision in Figure 7.2, and asking the only apparently vague question 'How much dynamics is there in this box? – how much "punch", how much "oomph" is here?' Saying 'there are two particles of mass $m$' doesn't seem an adequate answer, because it doesn't account for the fact that the particles are moving towards each other. Different observers will disagree about which particle is moving, and how quickly, but there can be no doubt that the particles are going to hit each other, quite hard. As well as the two particles, there's a lot of kinetic energy in this box (Table 7.2 tells us that, in the centre-of-momentum frame, the KE of the incoming particles is $E_i - m_i = 2$, a significant fraction of the mass energy $m = 8$).

If we imagine that these balls of relativistic putty are each a mere eight grammes in mass, and reverting to physical units for the moment, the incoming particle in the lab frame, which is moving at a healthy $15/17 \times c$ (see Table 7.1), has an energy of $E = \gamma mc^2 = 17/8 \times (8 \times 10^{-3}\,\text{kg}) \times (3 \times 10^8\,\text{m s}^{-1})^2 = 1.53\,\text{PJ}$ (about a third of a megaton of TNT), only $720\,\text{TJ}$ of which is accounted for by the energy equivalent of the putty's 8 g of mass, and the rest of which, amounting to $810\,\text{TJ}$, is kinetic energy. That's a lot of energy, and it may not then be surprising that that energy of motion makes as important a 'dent' in spacetime as the energy represented by the particles' masses.

That is, there is more mass after this collision than before, but not more 'gravitational charge'. A gravity detector (that is, another nearby mass) would not notice a change from immediately before to immediately after this collision.[1]

---

[1] To be very precise, movement of masses like this would in principle generate gravitational waves, but that's a complication we should not indulge in at this particular point.

Thus, as we have seen above, it is not mass that is conserved in collisions, but 4-momentum (or energy-momentum), **P**. In General Relativity, it is not mass that is the source of gravitation, but 4-momentum.

Taylor & Wheeler (1992, ch. 7) is an excellent discussion of the puzzles here.

## 7.5  More Unit Fun: an Aside on Electron-volts

We're going to start off using physical units in this section, partly in order to show why they're inconvenient for one of relativity's most common application domains.

If an electron (mass $9.11 \times 10^{-31}$ kg) is moving close enough to the speed of light that it has a $\gamma = 150$, then from Eq. (7.6) we discover that it has a total energy of $1.23 \times 10^{-11}$ J. Is that a large number in this context? Certainly, it's clear that the joule isn't a convenient unit here.

If an electron is accelerated through a potential difference of one volt, then it acquires a kinetic energy of $E = qV = 1.602 \times 10^{-19}$ J. This is a convenient unit of energy in particle physics, where it is known as the *electron-volt*, written 'eV'. The electron-volt is nothing more than a name for a convenient quantity of energy, just as the jansky, at $10^{-26}$ W, is the name for a small quantity of power, convenient for radio astronomers. Using this unit, this electron has an energy of $76.8 \times 10^6$ eV, or 76.8 MeV. Does *that* count as a large number?

If this electron, and a positron moving at the same speed, were to collide with one another, the total energy in the CM frame would be 154 MeV. Now, a neutral pion ($\pi^0$) is a subatomic particle which has a rest mass of $2.40 \times 10^{-28}$ kg, which you can therefore confirm has a rest energy of 135 MeV. Thus there is enough energy in this electron–positron collision for the two incoming particles to annihilate and result (briefly) in a pion. In such a context as this – where we are asking if there is enough energy available to create new particles – we are much less interested in the mass of a particle than in its rest energy, and so it is much more natural to quote the particle's mass, in a data table, in terms of that rest energy. Keeping the units correct, we would therefore write that $m(\pi^0) = 135$ MeV$/c^2$, and similarly that the mass of the electron is $m_e = 0.511$ MeV$/c^2$: just as the convenient unit of energy is the eV, the convenient unit of mass is the 'eV$/c^2$', which you can confirm has the dimensions of mass.

Similarly, we might look at Eq. (7.9), which in physical units is

$$E^2 = p^2c^2 + m^2c^4,$$

and observe that $pc$ has the dimensions of energy, and thus can be expressed in eV, so that $eV/c$ is a useful unit for momentum.

As you will recall, in natural units the speed $c = 1$, and this extra factor of $c^2$ becomes redundant. Thus in natural units the units of mass are the *same* as the units of energy, and we habitually write $m_e = 0.511$ MeV.

## 7.6  Relativistic Force ⚠

Another way of thinking about this is through the idea of *relativistic force*. Possibly surprisingly, force is a less useful concept in relativistic dynamics, than it is in newtonian dynamics – it's more often useful to talk of momentum and energy, and indeed energy-momentum. However, there are a few useful things we can discover.

We can define force in the usual way, in terms of the rate of change of momentum. Defining the 4-vector

$$\mathbf{F} = \frac{d\mathbf{P}}{d\tau},$$

we can promptly discover that $\mathbf{F} = m\mathbf{A}$. Referring back to Eq. (6.16), and writing $\mathbf{f}$ for the spatial components of the 4-vector, $\mathbf{F}$, we discover that $\mathbf{f} = m\gamma(\dot{\gamma}\mathbf{v} + \gamma\mathbf{a})$ (observing that in the non-relativistic limit $\mathbf{f} = m\mathbf{a}$ reassures us that this quantity $\mathbf{F}$ does have at least something to do with the thing we are familiar with as 'force', and that we are justified in calling $\mathbf{f}$ the 'relativistic 3-force'). But this tells us that $\mathbf{f}$ is not parallel to $\mathbf{a}$, as we (and Newton) might expect.

What else might we expect? In newtonian dynamics, $\mathbf{f} \cdot d\mathbf{s}$ is the work done on a particle by a force which displaces it by an amount $d\mathbf{s}$ – what is the analogue in relativity? We can write $\mathbf{F} \cdot d\mathbf{R} = \mathbf{F} \cdot \mathbf{U}\,d\tau$, but

$$\mathbf{F} \cdot \mathbf{U} = m\frac{d\mathbf{P}}{d\tau} \cdot \mathbf{U} = \mathbf{P} \cdot \frac{d\mathbf{P}}{d\tau} = \frac{1}{2}\frac{d}{d\tau}(\mathbf{P} \cdot \mathbf{P}) = 0,$$

since we know that $\mathbf{P} \cdot \mathbf{P}$ is conserved. So, although in newtonian dynamics the effect of a force on a particle is to do 'work' on it, and so change its energy, the effect of a *relativistic* force $\mathbf{F}$ is *not* to change the particle's energy-momentum, but instead to simply 'rotate' the particle in Minkowski space – it changes the direction of the particle's velocity vector, $\mathbf{U}$, but not its (invariant) length.

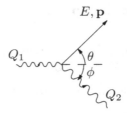

**Figure 7.3** Compton scattering: a photon being scattered from a charged particle.

Of course this *does* increase the particle's energy and momentum separately. If the particle starts off at rest in some frame, then after the force has been applied it will have $P^0 = m\gamma$, and $\mathbf{p} = \gamma m\mathbf{v}$. However, the time and space components of the momentum $\mathbf{P}$ are related by Eq. (7.8), $E^2 - \mathbf{p} \cdot \mathbf{p} = \mathbf{P} \cdot \mathbf{P} = m^2$, the mass of the particle, which is unchanged by the application of the force. Because of the minus sign in this expression, the time and space components of the momentum vector can increase without the magnitude of the vector increasing, and this is closely analogous to the role of the minus sign in our analysis of the twins paradox, in Section 5.7.1: there, the (travelling) twin who took the indirect route into the future travelled through *less* spacetime than the stay-at-home who took the direct route; here, the accelerated vector has a momentum with larger (frame-dependent) components, which nonetheless has the same length as it started with.

In both cases, the unexpected conclusions follow ultimately from the observation that Minkowski space is not euclidean space, and that in Minkowski space, instead of Pythagoras's theorem, we have Eq. (6.6).

## 7.7 An Example: Compton Scattering

As a further example of a relativistic collision, we can examine the collision between a photon – a quantum of light energy – and an electron. This is *Compton scattering*, and is not the same as the classical Thomson scattering of light by electrons: that is a purely electromagnetic effect whereas this is an inherently relativistic and quantum-mechanical effect, where we are treating both the electron and the incoming light as relativistic particles. See also French (1968, pp. 194–196).

The collision is as shown in Figure 7.3. An incoming photon strikes a stationary electron and both recoil. The incoming photon has energy $Q_1 = hf_1 = h/\lambda_1$ and the outgoing one $Q_2 = hf_2 = h/\lambda_2$; the outgoing electron has energy $E$, spatial momentum $\mathbf{p}$, and mass $m$. The four momentum 4-vectors are therefore

$$\mathbf{P}_{1e} = (m, 0, 0, 0) \quad \mathbf{P}_{2e} = (E, p\cos\theta, p\sin\theta, 0)$$
$$\mathbf{P}_{1\gamma} = (Q_1, Q_1, 0, 0) \quad \mathbf{P}_{2\gamma} = (Q_2, Q_2\cos\phi, Q_2\sin\phi, 0),$$

where subscripts 1 and 2 denote momenta before and after the collision, respectively.

Momentum conservation implies $\mathbf{P}_{1e} + \mathbf{P}_{1\gamma} = \mathbf{P}_{2e} + \mathbf{P}_{2\gamma}$. Comparing components, we find

$$Q_1 + m = Q_2 + E \tag{7.14a}$$
$$Q_1 = p\cos\theta + Q_2\cos\phi \tag{7.14b}$$
$$0 = p\sin\theta + Q_2\sin\phi. \tag{7.14c}$$

Writing Eq. (7.14b) as $Q_1 - Q_2\cos\phi = p\cos\theta$, and squaring and adding this and Eq. (7.14c), we obtain

$$(Q_1 - Q_2\cos\phi)^2 + (-Q_2\sin\phi)^2 = p^2. \tag{7.15}$$

Using the relation $p^2 = E^2 - m^2$ in this equation, then substituting for $E$ from Eq. (7.14a) and rearranging, we find

$$Q_1 Q_2 (1 - \cos\phi) = (Q_1 - Q_2)m, \tag{7.16}$$

which, on dividing through by $Q_1 Q_2 m$, gives

$$\frac{1}{Q_2} - \frac{1}{Q_1} = \frac{1 - \cos\phi}{m} \tag{7.17}$$

or, in terms of wavelength,

$$\lambda_2 - \lambda_1 = \frac{h}{m}(1 - \cos\phi). \tag{7.18}$$

We can see that the scattered photon will have a longer wavelength (smaller energy) than the incident one. Note also that this is not any type of Doppler effect: the photon has a different energy not because we are observing it in any sort of moving frame, but has a lower energy because it has had to give some up to the electron.

## 7.8 Not The End

That is all we have time and space for. Over the course of this text, we have examined the axioms on which SR is based, derived the immediate consequences of these axioms in the form of the Lorentz transformation equations, and finally examined how our understanding of dynamics has to be transformed within the context of relativity.

It is also just about all of the Special Relativity that you will need in your future study of physics. It is also, possibly surprisingly, just about all the SR that there is – there is no 'Advanced Special Relativity' text to move on to. More advanced courses that use SR depend on it as a foundation: they build on it but do not extend it. Even fascinating texts such as Schwarz & Schwarz (2004) use it as a starting point for excursions into neighbouring areas of physics and maths, and Gourgoulhon (2013) covers much the same ground with more sophisticated mathematical tools, but they do not discover *more* of it, as such.

In Appendix A, I provide a compact introduction to General Relativity, which builds directly on what we have learned about Special Relativity here, but necessarily without the extra maths which is required for a full understanding.

Has it been worth it? While you will only *very* rarely have to calculate the speed of clocks on passing spaceships, and rather less often on passing trains, you may well have to calculate the speed of a clock on a passing pion. And stepping back from calculation, the physical world picture we have looked at in this book is what underlies areas of physics as diverse as Relativistic Quantum Mechanics, Quantum Field Theory (underlying particle physics), and General Relativity, which are simply unintelligible otherwise. Even if you do not go on to study those areas, you have come to have a more fundamental view of the universe than you had from your intuitive understanding of Newton's physics, and gained access to one of the twentieth century's most radical intellectual adventures.

Enjoy!

## Exercises

### Exercise 7.1 (§7.1)

Recall the definition of the momentum 4-vector:

$$\mathbf{P} = m\mathbf{U} = m\gamma(1, \mathbf{v}).$$

What are the time- and $x$-components of the momentum of a particle of mass 2 kg moving with speed $v = 3/5$ (so $\gamma(v) = 5/4$) along the $x$-axis?

1. $P^0 = 5/2, P^1 = 3/2$

2. $P^0 = 5/2, P^1 = 6/5$
3. $P^0 = 1, P^1 = 3/5$
4. $P^0 = 5/4, P^1 = 3/2$                                    $[\,d^-\,]$

## Exercise 7.2 (§7.1)

We were a bit sloppy in Exercise 7.1. What are the units of $P^0$ and $P^1$?

1. $\mathrm{kg\,m\,s^{-1}}$ and $\mathrm{kg\,m\,s^{-1}}$
2. kg and $\mathrm{kg\,m\,s^{-1}}$
3. kg and kg                                                $[\,d^-\,]$

## Exercise 7.3 (§7.1)

The luminosity of the Sun is $L_\odot = 3.86 \times 10^{26}$ W; this is powered by nuclear fusion. How much mass does it consume, in fusion reactions, per second?

## Exercise 7.4 (§7.3)

Conserved quantities: are the following statements true or false?

1. Momentum is conserved in collisions.
2. Energy is conserved in collisions.
3. Invariant mass $(m^2)$ is conserved in collisions.          $[\,d^-\,]$

## Exercise 7.5 (§7.3)

Invariant quantities: are the following statements true or false?

1. Momentum is an invariant quantity.
2. Energy is an invariant quantity.
3. Mass $(m^2)$ is an invariant quantity.                      $[\,d^-\,]$

## Exercise 7.6 (§7.3.1)

A fast-moving particle of mass $m_1 = 1\,\mu\mathrm{g}$ is moving at speed $v = 0.5$ when it hits a stationary particle of mass $m_2 = 1000\,\mu\mathrm{g}$. What is the speed of the centre-of-momentum frame? What are the two particles' energies and momenta in the lab frame (that is, the one in which particle 2 is stationary), and in the CM frame?

Give the answers in both natural units and physical units (SI) (remember Exercise 4.1).

After you have read Section 7.5, additionally give the answers, including the two masses, in physical units in terms of eV.

# Appendix A
## An Overview of General Relativity

Mathematics, physics, chemistry, astronomy, march in one front. Whichever lags behind is drawn after. Whichever hastens ahead helps on the others. The closest solidarity exists between astronomy and the whole circle of exact science.
*Karl Schwarzschild, in 1913, quoted in Eddington's obituary of him (1917)*

Although, as I mentioned at the end of the last chapter, we have now covered a substantial fraction of what there is to say about *Special* Relativity, this is very far from all we can say about relativity in general. Special Relativity discusses the special case of motion within, and transformation between, inertial frames moving with constant velocity relative to each other. If we relax this restriction, and ask about transformations between arbitrary frames (i.e., between arbitrary coordinate systems), which may be accelerating with respect to each other or moving under the influence of gravity, then we are asking the questions of General Relativity – GR.

General Relativity is mathematically much more challenging than SR, and so the account below does give in to 'it can be shown that...' on more than one occasion, and it makes reference to mathematical technology such as differential geometry which we can talk about only schematically. However, the way we have covered SR in this course, with its emphasis on an axiomatic approach and on geometry, gives us a starting point from which we are able to do more than merely sketch out the landscape here.

You can find alternative explanations of some of the fundamental ideas – such as the Equivalence Principle, curvature, and cosmological expansion – online and in popular books. You can also learn something from the introductory chapters of GR textbooks, before they get properly started on the tensor calculus. I mention one or two such books in Section 1.8. In particular, I'll draw attention to Schutz (2009) and Rindler (2006); and also

145

Misner, Thorne & Wheeler (1973), which is justly famous, though it has a slightly idiosyncratic take on the subject. I emphasise, however, that these are all graduate texts, so that the level of maths required will quickly outstrip what I can prepare you for in this short chapter.

What is the problem that General Relativity is trying to solve?

We examine the problem of gravity in Section A.1, through a sequence of thought experiments, which bring out some immediate consequences of the ideas. The way we can think about the solution is by sticking with the theme of *geometry*, which we have laid so much stress on up to this point. In Section A.2 we learn about the ideas of the *metric* of a space, of the *curvature* of a space, and of *geodesics* through that space. Then, in Section A.3, we discover, in general terms, how we can put these ideas together to get a relativistic theory of gravity, in the form of *Einstein's equations* for the structure of spacetime around a gravitating mass. The mathematical details are beyond the scope of this course, but we can understand the shape of the ideas. Finally, in Section A.4, we look briefly at some solutions to Einstein's equations, and discover the continuity with Newton's theory of gravity.

## A.1  Some Thought Experiments on Gravitation

In the text so far, we have dealt with gravity by ignoring it or, where we couldn't do that, for example because we were talking of throwing balls or juggling, by quietly presuming it to be an uncomplicated 'downward-pointing' force.

General Relativity – Einstein's theory of gravitation – adds further significance to the idea of the inertial frame. Indeed, at this point we slightly *redefine* what we mean by an inertial frame. In the context of GR, an inertial frame is a frame in which SR applies, and thus the frame in which the laws of nature take their corresponding simple form. This definition, crucially, applies *even in the presence of large masses* where (in newtonian terms) we would expect to find a gravitational force. The frames thus picked out are those which are in *free fall*, either because they are in deep space far from any masses, or because they are (attached to something that is) moving under the influence of 'gravitation' alone. I put 'gravitation' in scare-quotes because it is part of the point of GR to demote gravitation from its newtonian status as a distinct physical force to a status as a mathematical fiction – a conceptual convenience – which is real only in the same way that centrifugal force is real, as a way of explaining the physical results of a change of frame.

In this section, we work through a number of thought experiments, which one by one introduce some of the key ideas, and some of the key conclusions, of GR.

### A.1.1 Standing in a Rocket

The first step of that demotion is to observe that the force of gravitation (I'll omit the scare-quotes from now on) is strangely independent of the nature of the things that it acts upon. Imagine a frame sitting on the surface of the Earth, and in it a person, a bowl of petunias, and a radio, at some height above the ground: we discover that, when they are released, each of them will accelerate at the same rate towards the floor (Galileo is supposed to have demonstrated this same thing using the Tower of Pisa, careless of the health and safety of passers-by). Newton explains this by saying that the force of gravitation on each object is proportional to its gravitational mass (the gravitational 'charge', if you like); and the acceleration of each object, in response to that force, is proportional to its inertia, which is proportional to its inertial mass. Newton doesn't put it in those terms, of course, but he also fails to explain why the gravitational and inertial masses, which *a priori* have nothing to do with each other, turn out experimentally to be *exactly* proportional to each other, even though the person, the plant, the plant pot, and the radio broadcasting electromagnetic waves all exhibit very different physical properties.[1]

Einstein supposed that this was not a coincidence, and that there was a deep equivalence between acceleration and gravity (we will see later, in Section A.3, that the force of gravity we feel standing in one place is the result of us being accelerated away from the path we would have if we were

---

[1] Newton does mention that 'a bit of fine down and a piece of solid gold descend with equal velocity' in the *General Scholium* to the second edition (1713) of his *Principia* (see for example https://isaac-newton.org/general-scholium/), but that text is better known for Newton's acknowledgement that, although he has described gravity, he cannot explain its source: 'Hitherto we have explaind the phaenomena of the heavens and of our sea, by the power of Gravity, but have not yet assignd the cause of this power. [...But] I frame no hypotheses.' Part of the problem was that the expectations of the time presumed that all forces were transmitted by contact, so that the idea of a pervasive gravitational field was contrary to all intuition, and a much more radical conceptual innovation than we might expect. Indeed it was possibly only Newton's occult and alchemical interests – research into which he pursued in parallel with his scientific work – that allowed him to think in such terms.

**Figure A.1** Objects in a spaceship.

in free fall). He raised this to the status of a postulate: the Equivalence Principle.

Putting Galileo to one side for a moment, imagine this same frame – or, for the sake of concreteness and the containment of a breathable atmosphere, a spacecraft – isolated in space (Figure A.1). Since spacecraft, observer, petunias and radio are all equally floating in space, none will move with respect to another (or, if they are initially moving, they will continue to move with constant relative velocity). That is, Newton's laws of motion work in their simple form in this frame, which we can therefore identify as an inertial frame.

If, now, we turn on the spacecraft's engines (Figure A.2), then the spacecraft will accelerate, but the objects within it will not, until the spacecraft collides with them, and starts to accelerate them by pushing them with what we will at that point decide to call the cabin floor. Before they hit the floor, they will move towards the floor at an increasing speed, from the point of view of someone in the box.[2] Crucially – and, from this point of view, obviously – the sequence of events here is independent of the details of the structure of the ceramic plant pot, the biology of the observer and the petunias, and the electronic intricacies of the radio. If the spacecraft continues to accelerate at, say, $9.81 \, \mathrm{m\,s^{-2}}$, then the objects now firmly on the cabin floor will experience a continuous force of one standard Earth gravity, and observers within the cabin will find it difficult to tell whether they are in an accelerating spacecraft or in a uniform gravitational field.

In fact we can make the stronger statement – and this is another physical statement which has been verified to considerable precision in, for example, Eötvös-type experiments[3] – that the observers will find it *impossible* to tell

---

[2] By 'point of view' I mean 'as measured with respect to a reference frame fixed to the box', but such circumlocution can distract from the point that this is an *observation* we're talking about – we can see this happening.

[3] These experiments use a torsion balance to directly confirm that masses of different materials experience the same gravitational force; for a review, see Adelberger et al. (2009).

**Figure A.2** Objects on the floor of an accelerating spaceship.

the difference between acceleration and uniform gravitation, and we can elevate this remark to a physical principle.

*The Equivalence Principle (EP):* Uniform gravitational fields are equivalent to frames that accelerate uniformly relative to inertial frames.

(A.1)

The Equivalence Principle is a *physical statement* rather than a mathematical one. By a 'physical statement' I mean a statement that picks out one of multiple mathematically consistent possibilities, and says that *this one* is the one which matches our universe. Mathematically, we *could* have a universe in which the EP did not hold, or in which the galilean transformation works for all speeds and the speed of light is infinite; but we don't.

That is, the EP (or at least this version of it) has now explained why all objects fall at the same rate in a gravitational field.

Note that I am at this point carefully *not* using the word 'accelerate' for the change in speed of the objects in the box with respect to that frame. We reserve that word for the physical phenomenon measured by an accelerometer, and the result of a real force, and try to *avoid* using it (not, I fear, always successfully) to refer to the second derivative of a position. Depending on the coordinate system, the one does not always imply the other, as we will see shortly.

## A.1.2 The Falling Lift

Recall from Special Relativity that we may define an inertial frame to be one in which Newton's laws hold, so that particles that are not acted on by an external force move in straight lines at a constant velocity. In Misner, Thorne and Wheeler's words, inertial frames and their time coordinates are defined so that motion looks simple. This is so if we are in a box far

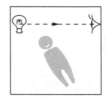

**Figure A.3** An observer in a floating box.

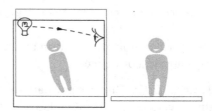

**Figure A.4** Falling down a lift-shaft.

away from any gravitational forces, and so we may identify that as a *local inertial frame* (LIF; we will see the significance of the word 'local' below). Another way of removing gravitational forces, less extreme than going into deep space, is to put ourselves in free fall. Einstein asserted that these two situations are indeed fully equivalent, and defined an inertial frame, firstly, as one in free fall, and declared, secondly, that just as all inertial frames in SR are equivalent, all (redefined) inertial frames in GR are equivalent, too.

> *Equivalence Principle, version 2:*   All local, freely falling, non-rotating laboratories (i.e., LIFs) are fully equivalent for the performance of all physical experiments.                                                    (A.2)

Imagine being in a box floating freely in space, and shining a torch horizontally across it from one wall to the other (Figure A.3). Where will the beam end up? Obviously, the beam will end up at a point on the wall directly opposite the torch. There's nothing exotic about this. The EP, however, tells us that the *same* must happen for a box in free fall, Figure A.4, left. That is, a person inside a falling lift would observe the torch beam to end up level with the point at which it was emitted, in the (inertial) frame of the lift. This is a straightforward and unsurprising use of the EP.

How would this appear to someone watching the lift fall? Since the light takes a finite time to cross the lift cabin, the spot on the wall where it strikes will have dropped some finite (though small) distance, and so will be lower

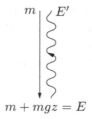

**Figure A.5** A falling mass and rising photon.

than the point of emission, in the frame of someone watching this from a position of safety (Figure A.4, right). That is, this non-free-fall observer will measure the light's path as being curved in the gravitational field. The EP forces us to conclude that even massless light is affected by gravity.

> Note that it is useful to be clear exactly where the EP enters this argument. It lets us go from a scenario we are confident we can fully analyse – namely the box far away from gravitating matter – to a scenario we might be more hesitant about – namely motion under the influence of gravity. We will use a version of the EP again in Section A.3, to move from a context we understand – physics in SR – to one we initially don't – physics in GR.

### A.1.3 Gravitational Redshift

Next, imagine dropping a particle of mass $m$ through a distance $z$. The particle starts off with energy $m$ ($E = mc^2$, with $c = 1$), and ends up with energy $E = m + mgz$ (see Figure A.5). Now convert all of this energy into a single photon of energy $E$, and send it up towards the original position. It reaches there with energy $E'$, which we convert *back* into a particle.[4] Now, either we have invented a perpetual motion machine, or else $E' = m$:

$$E' = m = \frac{E}{1 + gz},\qquad\text{(A.3)}$$

and we discover that a photon loses energy as a necessary consequence of climbing a distance $z$ through a gravitational field, and as a consequence of our demand that energy be conserved.

Let's look more closely at this photon, and take it to be climbing through a positive distance $\Delta r$. As Planck and Einstein learned, the photon's energy

---

[4] As described, this is kinematically impossible, since we cannot do this and conserve momentum, but we can imagine sending distinct particles back and forth, conserving just energy; this would have an equivalent effect, but be more intricate to describe precisely.

is $E = hf$, where $h$ is Planck's constant and $f$ the photon's frequency, in hertz. From Eq. (A.3), we can write $E = E'(1 + g\Delta r)$, and writing $E' = hf' = h(f + \Delta f)$, we discover that $hf = h(f + \Delta f)(1 + g\Delta r)$. Taking the differences to be infinitesimal, and discarding terms in $O(dr\,df)$, we then find

$$\frac{df}{f} = -g\,dr.$$

Now we know that this acceleration $g$ is the result of a gravitational force $F = mg$, or $F = m\nabla\phi$, where $\phi$ is Newton's gravitational potential. If we suppose that this potential is that of a body such as the Earth, and $r$ is a radial coordinate, then $\phi = \phi(r)$, and $g = \nabla\phi = d\phi/dr$. Thus $gdr = d\phi$, and

$$\frac{df}{f} = -d\phi.$$

We can integrate this between, say, radii $r_0$ and $r_1$ to find

$$\frac{f_1}{f_0} = \frac{e^{-\phi_1}}{e^{-\phi_0}}, \tag{A.4}$$

where we can if we wish choose the constant of integration to set either one of the potentials to zero.

The frequency of radiation is of course inversely related to its period, so we can repurpose Eq. (A.4) to obtain an expression for the change in periods $\Delta t$ of radiation at the two potentials:

$$\frac{\Delta t_0}{\Delta t_1} = \frac{e^{-\phi_1}}{e^{-\phi_0}}. \tag{A.5}$$

This frequency shift, corresponding to an energy loss, is termed *gravitational redshift*,[5] and it has been confirmed experimentally, in the 'Pound–Rebka experiments' of 1959–64,[6] which were able to detect the frequency

---

[5] This is also sometimes referred to as 'gravitational Doppler shift', but inaccurately, since it is not a consequence of relative motion, and so has nothing to do with the Doppler shift you are familiar with. Also, if the photon is descending, it is blue-shifted, so that 'gravitational frequency shift' would be a better term. But 'redshift' is conventional.

[6] Multiple research groups, including Pound's in Harvard, USA, and John Paul Schiffer's in Harwell, UK, realised in 1959 that they could use the Mössbauer effect (discovered in 1958) to make this measurement, and published data-less announcements of their plans to do so. Schiffer's group were the first to publish experimental results, in February 1960, followed only a few weeks later by the Harvard group's first publication; both groups announced a detection of the redshift, within experimental error. The Harwell group's results were, however, arguably the result of a systematic chemical effect which supplied or amplified the redshift result, and only Pound's group seems to have gone on to refine the measurement in a sequence of publications in the succeeding years. It is not unreasonable, therefore, that

changes of gamma rays falling a mere 22.5 m, in which the factor $gz/c^2$ in Eq. (A.3)[7] is only $2.5 \times 10^{-15}$.

Light, it seems, can tell us about the gravitational field it moves through.

## A.2 Geometry

To get from here to GR we need to answer three questions: (i) what do we mean by 'the geometry of a space'? (ii) how do we describe geometry mathematically? and (iii) how do we make the link from geometry to gravity?

Geometry is the mathematical discussion of shapes. We tend to think of this as a fixed thing, which we learned about in school, and haven't had to think much more about since. Euclid famously established an axiomatic basis for geometry, setting out a small number of axioms, or postulates, on the basis of which a large number of geometrical theorems can be proved. The space he described, with the fairly straightforward extension from the plane to higher dimensions, is referred to as euclidean space. In this space, the internal angles of a triangle add up to 180°, parallel lines never meet, the circumference of a circle is $\pi$ times its diameter, and Pythagoras's theorem gives the distance between points.

That was thought to be the end of the matter for a very long time.

One of Euclid's postulates is that there exist parallel lines – lines which never cross each other. This postulate can be expressed in a variety of ways, but it also seems either to be self-evident, or to be such a trivial addition to the other postulates that it must surely be provable from them; and for more than 2000 years after Euclid, mathematicians tried to do exactly that. It was only in the nineteenth century that the mathematical community discovered that one can consistently discuss geometry *without* the parallel postulate, or with alternative postulates such as the demand that an arbitrary line in the space has zero parallel lines, or that it has more than one. These alternative postulates lead to different conclusions about, for example, the sum of the

---

these are memorialised as the 'Pound–Rebka' measurements, but the episode usefully illustrates the contingency of experimental races. For a history of the episode, including references to the various original papers, see Hentschel (1996). The race, the pre-announcements, and the science-by-press-release, are also discussed rather unflatteringly in Reif (1961). Further, see Vessot et al. (1980) for a description of the 1976 Gravity Probe A experiment, which measured the effect between the ground and a space-based instrument at an altitude of 10 000 km.

[7] Where did the $c^2$ come from? Equation (A.3) is written in units where $c = 1$, as usual; in order to get a numerical value from '22.5 m', we must examine the dimensions of the expression to discover that the denominator, in physical units, must be $1 + gz/c^2$.

internal angles of a triangle, or the ratio of a circumference to a diameter. So when we talk about 'the geometry of a space', we mean by 'space' a set of points such as a plane or a volume, or Minkowski space; and by that space's 'geometry' we mean the set of 'rules' for distances, angles, and so on. If we use Euclid's rules, we get euclidean geometry; if we want to talk about Minkowski space, we have very similar rules, but with a definition of 'distance' which we have seen is different from the pythagorean distance.

But before we can talk about distance, we have to step back and think a little more clearly about coordinates.

## A.2.1  Coordinates

Where are you? What time is it?

At this stage in your physics education, you are comfortable with the idea of describing the positions of things, and their movement, by giving them *coordinates*, and talking about how those coordinates change. Those coordinates are a mathematical fiction, of course, but it's useful to stress how completely arbitrary those coordinates are.

In addressing one or other physics problem, we are taught to make things easy for ourselves by choosing cartesian coordinates, or polar coordinates, or one of the various other more exotic alternatives. Our choice here will have an effect on, for example, how we write down velocities, and the form of the gradient operator that we must use. Two key points are (i) the coordinates we use are a *choice*, and (ii) they are not physically significant.

The way we describe geometry mathematically (or at least the way which is of most relevance to GR) is using *differential geometry*. Unfortunately, this subject is too mathematically challenging to go into at the level of this text. What I can do instead is to give you a sense of the key ideas, in just enough depth that we can understand how they are used to make the link to gravity.

A *coordinate system* is a way of systematically drawing grid-lines throughout a region of spacetime, which is curved or flat, and thus attaching coordinates to each event. These coordinates *don't have any intrinsic meaning* – they're just labels attached to points in space and time, and the difference in coordinates between two events doesn't tell you anything immediate about how far apart they are. Also, we discover that, in our universe, we will always need three space and one time dimension to fully and non-redundantly locate a point.

This isn't too hard to think about for space dimensions: we can of course imagine grid-lines drawn on the ground, and perhaps imagine some arrangement with balloons or helicopter drones, with numeric coordinates painted on the sides of them, and perhaps observers sitting in some of them. It's a little harder to think about for time. For Newton, this was easy: the newtonian picture is of a sequence of three-dimensional 'snapshots', each of which has a time-value attached to it. There's some clear arbitrariness to the timescale, in terms of how fast the clock ticks, or where the zero of time is, but there's equally well a clear association between each snapshot and 'the time'. A brief reflection on Chapter 4 will tell us that things aren't going to be so simple in a spacetime built on Minkowski space.

Probably the most straightforward way to define a time coordinate is to decide that the time coordinate of an event is the number showing on the watch of an observer located there. Or, more generally, that each one of our balloons or drones has a clock attached to it, which ticks in some systematic way, and 'the time' is the number on that clock face. It might be that these clocks run differently on different drones, so that it's not a given that they all tick 'together'.

### A.2.2 The Metric of Euclidean Space

In Chapter 5 we derived the Lorentz transformation by demanding that the interval $s^2$ is invariant under that transformation. Our eventual definition of the interval is in Eq. (4.4), and as such we can now recognise this as the first appearance of the *metric*. Quite generally, we regard the metric of a space as the definition of length in that space. What does that mean?

The metric in euclidean space is

$$ds^2 = dx^2 + dy^2, \tag{A.6}$$

which you will recognise as a differential form of Pythagoras's theorem. This definition doesn't pin down everything – it doesn't say what length units we use, nor where the origin is – but it does give a systematic recipe for turning coordinate differences $dx$ and $dy$ into a length $ds$.

How would we use this to calculate the distance between two points $P_0 = (x_1, y)$ and $P_1 = (x_2, y)$? That's not hard: we simply calculate

$$l = \int_{P_0}^{P_1} ds = \int_{P_0}^{P_1} \sqrt{ds^2} = \int_{P_0}^{P_1} \sqrt{dx^2 + dy^2}$$

$$= \int_{x_0}^{x_1} \sqrt{dx^2}, \qquad \text{since } dy = 0 \text{ along this path}$$

$$= \int_{x_0}^{x_1} dx = x_1 - x_0.$$

That's not unexpected.

However, we can also describe points in the plane using polar coordinates, $(r, \theta)$. The distance element – that is, the *metric* for polar coordinates – is

$$ds^2 = dr^2 + r^2 d\theta^2, \tag{A.7}$$

which you have also probably seen before, in one guise or another. We could also use this metric to obtain the distance between points $P_0$ and $P_1$. This is a slightly fiddly calculation, because $(r, \theta)$ would be a poor choice of coordinates to use, but the end result would be the same distance: the two metrics are defined in relation to different coordinates, and so they provide different recipes for going from coordinate differences to distances in the space; but the space they are describing – flat euclidean space in this case – is the same, so the distances they produce must be identical.

That is, the metric is describing a measurable property of a space – the length of a path through that space – but each version of that metric works with a particular set of coordinates. The set of coordinates is otherwise (largely) arbitrary. Dealing with that arbitrariness, and learning how to talk about distance, and differentiation, in the presence of that arbitrariness, is what differential geometry is all about.

### A.2.3  The Metric of a Non-euclidean Space: Curvature

As an example of a space which is *not* euclidean, let us look at coordinates on the surface of a sphere. Here, the line element on the surface of the sphere (in terms of co-latitude $\phi$) is

$$ds^2 = R^2(d\theta^2 + \sin^2 \theta d\phi^2). \tag{A.8}$$

Here, we are talking about the *two-dimensional surface* (2-d) of a sphere of radius $R$, and are not referring to the volume element of three-dimensional (3-d) space in spherical polar coordinates.

It is not obvious, just from looking at this, that it describes a non-euclidean space. To learn more, we must at least mention 'spherical trigonometry'. This is the study of triangles on the surface of a sphere, which navigators or astronomers learn about, since they want to be able to calculate distances or angles on the surface of the ocean, or on the sphere of the sky.

In Figure A.6 we see a triangle on a sphere of radius $R$, with sides of length $a$, $b$ and $c$, and internal angles $A$, $B$ and $C$ (note that most treatments of spherical trigonometry will quote expressions with a nominal sphere radius of $R = 1$).

On the surface of a sphere, these internal angles of the triangle add up

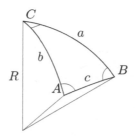

**Figure A.6** A spherical triangle.

to a value of $s = A + B + C$ which is greater than $\pi$ radians, and if we draw a circle on the surface, consisting of all of the points a distance $L$ from the pole, we would find its circumference to be less than $2\pi L$ (for example, the equator is fairly obviously shorter than $2\pi$ times the distance from the pole to the equator). That is, these basic geometrical rules are not the same as they are in euclidean space.

Further, it may be clear that this excess $s - \pi$ depends on $R$; and specifically, as $R$ gets larger, this excess gets smaller: that is, the geometrical features of the space depend on a parameter, $R$, which is related to a property called the *curvature* of the space. Specifically, because of how it is defined, the curvature of this space is $1/R$, and as $R$ gets larger, and the surface becomes locally more like flat euclidean space, the curvature tends to zero. In this example, the 'curvature' is simply related to the radius of the sphere, but the notion is more general.

A key thing, here, is that if we lived on this surface, we could draw a triangle on the 2-d surface, make careful measurements of it, and discover this excess, and thus measure the parameter $R$ which appears in the metric, without knowing anything about the 3-d volume that the surface is embedded within. That is, the curvature is a property of the (2-d) space which is detectable by *local measurements* within the surface, and it is not, for example, an artefact of our perception of the surface in 3-d space.

In the case of the surface of a sphere, the curvature is the same over the entire space. If you think of the surface of an ellipsoid, in contrast, you can see that the curvature can vary as you move around the space. The curvature in these two cases is positive.

In contrast, a hyperbolic paraboloid, as in Figure A.7, is a surface with constant *negative* curvature (which therefore doesn't have such a straightforward relationship to anything like a radius). On this surface, the internal angles of triangles add up to less than $\pi$, and the circumferences of circles

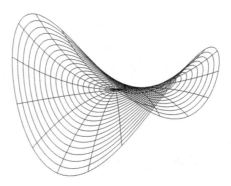

**Figure A.7** A hyperbolic paraboloid.

are greater than $\pi$ times the diameter.

The examples above have illustrated curvature in 2-d surfaces. The same general ideas carry over to surfaces of higher dimensions, even if they then become harder to visualise: a 3-d space can be similarly curved or flat. The volume of a sphere in three euclidean dimensions is, as you'll recall, $4\pi R^3/3$, but a sphere in a curved space would have a volume greater or smaller than this, depending on the sign of the curvature.

The case we are most interested in, of course, is that of *spacetime*, which is a space (in the mathematical sense) with one time and three spatial dimensions. Minkowski space, as we have seen it so far, is the special case of a spacetime of constant zero curvature, but, just as with the ellipsoid, the curvature can vary over the spacetime.

### A.2.4  Geodesics

When we draw a triangle in euclidean space, we obviously connect the three vertices with straight lines. The lines joining the vertices in the triangles of Section A.2.3 are not straight lines in 3-d space, but they *are* the nearest thing to a 'straight line' on the 2-d surface of the sphere (they are 'great circles'). This is telling us that we might have to generalise our notion of 'straight line' when talking about non-euclidean geometries. The result of this generalisation is the *geodesic*.

There are a couple of ways we can think of geodesics. One is to see them as the shortest distance between two points in the space. This works for euclidean space, and for the great circles on the surface of a sphere, but if you think back to Section 5.7.1, you might recall that I mentioned in passing that the straight line followed by the non-travelling twin was in fact the

**Figure A.8** Walking forwards on a sphere.

**Figure A.9** A trivial geodesic in Minkowski space.

*longest* distance between two points in Minkowski space. We can think of geodesics instead as the *path of extremal length* between two points (and see also Exercise 5.8). A great circle between two points which goes 'the long way round' is the maximal distance between the points on the surface of the sphere, and is also a geodesic.

An alternative way of thinking about geodesics is to conceive of them as the path you trace out in a space if you 'walk forwards'. If, in flat euclidean space you walk, cycle, or drive straight ahead then you will trace out a straight line; if you do the same on the surface of a sphere, you will trace out a great circle. This definition can be made more precise, but the key thing is that it is a *local* definition – you don't have to know the overall structure of the space in order to work out, or pace out, the geodesic.

What are the geodesics in Minkowski space?

The instruction 'walk forwards' in Minkowski space means 'move in a straight line at a constant speed', so the sequence of events associated with, say, a flickering light carried by an observer, traces out a geodesic in that space. We can also note – this will become important later – that for any observer moving in this way, there is an inertial frame in which they are at rest, and thus have a geodesic worldline lying along the $t$-axis (Figure A.9).

## A.3  Gravity

We now have the set of ideas, and the language, in which we can describe a relativistic theory of gravity – Einstein's General Relativity.

### A.3.1  How Matter Moves: Geodesics

Our first question is to ask: if you are moving under the influence of gravity, what *direction* are you moving in?

For that, we can return to the EP, quoted above as Statement A.2. This is similar to our statement about the universal equivalence of inertial frames, back in Section 2.1, but it is more general in that, by expanding the set of equivalent frames to 'all freely falling non-rotating' frames – which in the context of GR we refer to as 'inertial frames' – it has made it possible to start to talk about gravity as well.

We can make a still more restrictive statement and say

*Equivalence Principle, version 3:*   Any physical law that can be expressed geometrically in SR has exactly the same form in a locally inertial frame of a curved spacetime.                         (A.9)

*This* is why we have been so concerned with geometry all the way through this text.

Imagine doing some physics experiments in a free-falling frame, within a spacetime which is not necessarily the simple one we have come to understand as Minkowski space. This free-falling frame can be regarded as an inertial frame, within which SR tells us the rules. From Chapter 7 we know that $\mathbf{F} = d\mathbf{P}/d\tau$ in this inertial frame. What this final version of the EP is telling us is that, even if the spacetime we are falling within is significantly more complicated (i.e., curved) than Minkowski space, *this curvature does not change the physical laws*; there is no 'curvature coupling'. We could imagine the law $\mathbf{F} = d\mathbf{P}/d\tau$ acquiring some extra terms due to the curvature – there's no mathematical reason why it shouldn't – but it doesn't. That's a profound physical statement.

This tells us that we *already* understand the physics of how things move in the context of GR – particles move along the $t$-axis of their local inertial frame, and according to the force laws of Chapter 7. But at this point we cannot translate that understanding into the other frames that we care about, such as ones that are accelerating, or which are otherwise not free-falling (for example because they are stationary on the surface of the gravitating Earth). Since we saw above that our normal motion within an SR inertial frame is along a geodesic in Minkowski space, this EP tells us that this is true in GR as well, so that our motion in the spacetime of GR is to follow geodesics in that space. That is, the effect of the curvature of spacetime is to define the geodesics along which free-falling particles move. This is usually summed up by the slogan:

Spacetime tells matter how to move…

This discussion is about free-falling under gravity. How does this tell us something new about our familiar gravitational experience of standing on the ground? Whether we are standing on the ground or sitting in a chair, or holding desperately onto a trapeze, what we are *not* doing is falling towards the centre of the Earth. But if we were to compare ourselves to a ghostly alternative self which is indeed falling, we would find the gap between us increasing at a constant 9.81 m s$^{-2}$, and see ourselves being accelerated away from that falling figure by the force transmitted through our feet or arms, ultimately from the ground. That is, the 'force of gravity' that we feel is merely the acceleration away from our otherwise free-fall trajectory.

The conceptual material in this section is discussed in many GR text-books, without that discussion requiring much mathematical apparatus, so that it can be generally intelligible. For example, see Schutz (2009, §7,1) and Rindler (2006, ch. 1 and to some extent §8.4) for details. The various 'versions' of the EP quoted here have subtly but significantly different physical content (Rindler is good on the distinctions), but I have treated them as mere rephrasings for our present purpose, and referred to them all as just 'the Equivalence Principle'.

### A.3.2  How Space Curves: Einstein's Equations

We can at this point draw an analogy with newtonian gravitation, in which the gravitational force is the gradient of the gravitational potential, $\phi(\mathbf{r})$, and that potential is given by Poisson's equation with mass density as the source term:

$$\nabla^2 \phi = 4\pi G \rho(\mathbf{r}).$$

The mass on the right-hand side constrains the second derivative of the potential. We can solve this equation to find the potential $\phi$, and its gradient $\mathbf{f} = \nabla\phi$, which gives the gravitational force at a point. Einstein's equations for the gravitational field are broadly analogous to this.

You learned in Chapter 7 that (rest) mass is a frame-invariant quantity, but you also learned in Section 7.4 that, when things are moving at high speed, there is more energy-momentum in a system than is accounted for by the obvious mass, and that while the amount of mass is not conserved in a collision, the amount of energy-momentum is. It therefore makes sense that it is the energy-momentum that appears on the right-hand side of our relativistic gravity equation, but in what form?

This appears in the form of the 'energy-momentum tensor', where a 'tensor' is a geometrical object with a family relationship to vectors. We can think of a vector as thing which will provide an answer to a 'one-vector question' – meaning a question which is phrased in terms of a single vector. For example a force vector is something which, when combined in an inner product with a displacement vector, gives the work done in that displacement. The strain tensor (for example) can answer a 'two-vector question': 'what is the component of force in *this* direction across a surface oriented in *that* direction?' For our present purposes, you don't need to know anything more about tensors than this: that they are as geometrical, and as frame-independent, as the 4-vectors we have already learned about.

What, then, appears on the left-hand side of our analogue of Poisson's equation? Just as that equation has the mass density constrain the second derivative of the potential, we can define a tensor, called the *Einstein tensor*, which is a combination of second derivatives of the metric.

Through a sequence of physically motivated but nonetheless heuristic arguments, Einstein asserted in his 1915 paper that this tensor G was simply proportional to the energy-momentum tensor:

$$G = -\kappa T. \tag{A.10}$$

This is *Einstein's equation* for gravity, in which the constant $\kappa$ plays the same broad role as the gravitational constant $G$ in Newton's theory. This is a physical statement, rather than a mathematical deduction from anything, and it is a statement which is corroborated by experiment.

We have omitted a *lot* of mathematics here. This equation, simple though it might look here, represents a set of ten coupled second-order partial differential equations, and is fiendishly difficult to solve. It has been solved in a few cases, however, and we will look briefly at a couple of these in the next section.

The key thing to note, however, is that, just as with Poisson's equation, Einstein's equation represents a sequence of constraints. The distribution of energy and matter is represented by the energy-momentum tensor T. This constrains G, which in turn constrains the metric of the space. In less mathematical terms, the presence of a lump of matter changes the metric of the spacetime around it. Because the spacetime is curved, the path of the 'straight line' geodesic traced out by a particle in free fall is different from the geodesic worldline that the particle would trace out if the mass weren't there. We interpret that deflection of the worldline as the result of the 'gravity' of the central mass, so that we see the straight line in spacetime as the movement in space and time of a ball thrown through the air.

Newton explains gravity by saying that a mass creates a field around it, which results in a force on test particles, which is proportional to their mass and directed towards the centre. Einstein explains the same behaviour by saying that the effect of a mass is to change (the metric of) the spacetime around it. Nearby free-falling test particles 'know' nothing of the central mass, but simply move straight ahead along geodesics in their local inertial frame (Figure A.9). The shape of the spacetime, however, has the effect of causing the particle to move along a path which we describe as an 'orbit'. An orbit is simply a straight line in a curved space.

It is astonishing that Newton's and Einstein's models, which start from such different places, with such different motivations, nonetheless produce predictions for the behaviour of free-falling particles which so very precisely match each other, and match reality.

This relationship allows us to complete the other half of the famous slogan, originally framed by John Wheeler,

Spacetime tells matter how to move;
                                                  matter tells spacetime how to curve.

## A.4 Solutions of Einstein's Equation

At this point, we have Einstein's equations, which relate the metric for a spacetime to the energy-momentum within it. The process of *solving* these equations is, as I have mentioned, famously hard, and there is no general solution, but only a variety of solutions to various special cases. These special cases – for example, the case of an isolated mass, or a small mass – are typically approximations or idealisations of the more general case.

As well as exact solutions, it is possible to solve the equations numerically, and this is the locus of most of the currently mainstream astronomical research in GR.

Confining ourselves to the analytic solutions, however, the various special cases are important, and cover a large fraction of the astronomically important cases. We will look at a few of these in the remaining sections of this appendix.

I can go into very little mathematical detail here, but I hope that I can give you at least a flavour of how we would use the mathematical structures of SR, in the generalisation to gravity.

### A.4.1 The Weak-Field Solution, and Newton

The first special case to look at is the case where the gravitating mass is small, such as something around the mass of a planet. This creates what is known as the 'weak-field' approximation to the solution. We are also going to restrict ourselves to 'slow' movement, at speeds much less than that of light. Our goal is to describe the spacetime around this mass, and in doing so (as we have learned in the previous section) describe the effects of gravity there.

Our first step is to choose a coordinate system. With that choice made, our goal translates into finding a metric for our 'weak-field' spacetime, which is consistent with what we have learned about relativity and gravitation so far, and expressed using those coordinates.

In Section A.2.1, we saw that we can imagine a grid drawn on the ground, and extended upwards with drones or balloons. We can use our weak-field approximation immediately by deciding to draw this grid naively: we'll presume that the space is approximately flat, and draw a euclidean cartesian grid on it, or a spherical polar one, or whatever one we prefer. We'll write the metric of this flat space as $d\sigma^2$, and not care, for the present, whether this is $d\sigma^2 = dx^2 + dy^2 + dz^2$, or $d\sigma^2 = dr^2 + r^2(d\theta^2 + \sin^2\theta d\phi^2)$, or even just $d\sigma^2 = dx^2$ if we're caring only about movement along the $x$-axis.

For our time coordinate, we'll tell each one of our observers to station themself at a particular spatial coordinate, and issue each of them with an identical clock. This is a clock which will reliably tick out proper time (that is, the amount of time which has passed for it; think back, again, to the taffrail clock of Figure 1.4), but which is adjustable by the observer so that the time shown on the clock face advances systematically faster or slower than proper time. We will decide that the number shown on a particular clock face is the *coordinate time* at that location. At this point, any event in spacetime can be given one time and three space coordinates, and we can write down a metric, in these coordinates, as

$$ds^2 = (1 + A)dt^2 - d\sigma^2, \tag{A.11}$$

where $A$ is some position- and time-dependent correction, yet to be determined, which we will shortly discover is small, in the sense of $A \ll 1$. As you can see, at $A = 0$ this reduces to the familiar Minkowski metric.

In writing this down, we have also assumed that we are dealing with 'slow' movement, in the sense that $d\sigma^2 \ll dt^2$. That means that we have *not* included a factor $(1 + B)d\sigma^2$, where the 'small × small' term $Bd\sigma^2$ would be promptly discarded.

We have decided that the clocks can have their rates adjusted, but to what value? In Section 2.2.1 we saw how we can synchronise clocks in flat Minkowski space. Imagine doing a similar thing, but in the context of a spacetime with a non-flat metric. How do I, stationed at radius $r_{me}$, instruct an observer at radius $r$ how to set their clock? Let's decide that *my* clock is the reference standard, and ask 'what would we have to do to make the coordinate time difference between events at radius $r$ the same as that at $r_{me}$?' We have to slow down the clock's displayed time with respect to its proper time, by a factor which depends on the gravitational potential, $\phi$, at that point, as we saw in Eq. (A.5). Specifically, and arbitrarily taking $\phi(\mathbf{r}_{me}) = 0$, we can write

$$d\tau^2 = e^{2\phi(r)}dt^2, \tag{A.12}$$

to define the interval in *coordinate time* between two events, co-located at radius $r$, which are separated by a proper time $d\tau$. At this point we can rewrite Eq. (A.11) to get

$$ds^2 = e^{2\phi}dt^2 - d\sigma^2. \tag{A.13}$$

The newtonian potential $\phi = -GM/r$ has dimensions $L^2T^{-2}$, so that in physical units we would expect instead $\phi/c^2$ here: putting in standard physical-units values[8] for $GM_\odot$ and $r_\oplus$ gives $\phi/c^2 \approx 10^{-8}$, so that $\phi$, in Eq. (A.13), is indeed small, this leading term is very close to 1, and we can write this alternatively as

$$ds^2 = (1 + 2\phi)dt^2 - d\sigma^2, \tag{A.14}$$

omitting terms in $\phi^2$. This is a weak-field metric: a metric for the spacetime around a 'small' central mass. For small $\phi$, this is close to the metric of Eq. (4.4) or Eq. (6.7), but it is not quite the same. Our argument here has told us that the presence of the mass has changed the definition of distance in the spacetime around it.

How do particles move in this spacetime? We learned in Section A.3.1 that particles move along geodesics, and in Section A.2.4 that those geodesics are the lines of extremal length.

---

[8] Here, as is conventional in astronomical contexts, I use subscripts $\odot$ and $\oplus$, to refer to the Sun and Earth respectively.

If you have done some advanced classical mechanics (Hamill 2013) you will have learned that it is possible to obtain the equations of motion of an object in (for example) a gravitational field by extremising a quantity known as the 'lagrangian', formed from the difference between the kinetic and gravitational potential energy, or $L = mv^2/2 - m\phi$. It is possible to show that the demand that our geodesic extremises Eq. (A.14) produces the same constraint as this lagrangian formalism. In other words, the solution to our geometrical problem produces the same path $\mathbf{x}(t)$ as we would otherwise classically predict on the basis of motion under gravity.[9]

In other other words, we have rediscovered Newton's theory of gravity, as a 'weak-field' approximation to the behaviour of objects in free-falling inertial frames, as constrained by the Equivalence Principle. This is the first basic test of GR.

Notice that, in this argument, we have *not* used Einstein's equation, or any part of the argument of Section A.3.2 (though we briefly touched on the physics of Section A.3.1); that is, to be precise, this isn't really a weak-field 'solution'. Instead, we brought gravity into the argument by using what is essentially Newton's gravity: the argument was built directly on the idea of gravitational redshift (Section A.1.3), which in turn depends on the idea of a mass losing energy $mgh$ when it drops through a given distance. It is because we have imported gravity by this route that this argument is limited to low-speed (i.e., newtonian) motion within the field.

### A.4.2  The Weak-field Solution, and Einstein

An alternative route to a theory of weak-field gravity is to solve Einstein's equations in the same framework as the previous section – namely an isolated single mass – and the same low-mass approximation, but this time starting with Einstein's equation, rather than heuristically including Newton's gravitation. Unlike the previous section, this route necessarily involves the mathematics of differential geometry, so all we can do in this section is to indicate how the problem is set up, and then jump to the solution.

As before, we imagine a spacetime containing only a single gravitating mass. We assume that the mass is 'small' (which in this context means just about everything less dense than a neutron star). In this case, the spacetime must be approximately the same as that of empty space: it is approximately

---

[9] The argument here is elegant, but unfortunately very compressed. It is ultimately derived from Rindler (2006, §§9.1–9.4), where you can find the steps I have skipped, and a number of subtleties which I have elided.

Minkowski space. Put mathematically, that means that the metric of this spacetime will be approximately the metric of Eq. (4.4), plus some small perturbation.

Skipping the details, we discover that our invariant interval in this space – our metric – becomes

$$ds^2 = (1 + 2\phi)dt^2 - (1 - 2\phi)d\sigma^2, \qquad (A.15)$$

where $d\sigma^2 = dx^2 + dy^2 + dz^2$, and where the function $\phi(r)$ is a solution of

$$\nabla^2\phi = -\frac{\kappa}{2}\rho(r),$$

and $\kappa$ is a constant, the value of which we can fix by demanding compatibility with Newton's gravity. You can see that this reduces to Eq. (4.4) when $\phi$ is small. This is obviously also similar to Eq. (A.14) with the addition of the factor in front of $d\sigma^2$. It is an approximation for the case of small masses, but because of the way it has been obtained, this version is relativistic, and doesn't have the restriction to $d\sigma/dt \ll 1$.

We learned in Section A.3.1 that objects move along *geodesics*, so our next question is: what are the geodesics in this spacetime? We can discover that these geodesics are the result of the momentum 4-vector being constrained by the potential $\phi$, with the result that we rediscover

$$\mathbf{f} = -m\nabla\phi.$$

That is, as we would hope, the small mass limit of Einstein's theory of gravity is Newton's law of gravity – the law of gravity that we are familiar with. This demonstrates explicitly that Newton's theory of gravity is the non-relativistic limit of Einstein's.

### A.4.3 The Schwarzschild Solution

Now imagine the same configuration – a spacetime with a single point mass in it – but relax the constraint that the central mass be 'small'. We now have a harder problem to solve, but one which was nevertheless solved fairly rapidly. In 1916, only a few months after Einstein published his 1915 paper describing GR, Karl Schwarzschild published the first exact solution to Einstein's equation.

The so-called *Schwarzschild solution* is a metric in terms of the coordinates $(t, r, \theta, \chi)$, where $\theta$ and $\chi$ are the usual polar coordinates. In these coordinates, the metric is

$$ds^2 = \left(1 - \frac{R}{r}\right)dt^2 - \left(1 - \frac{R}{r}\right)^{-1}dr^2 - r^2 d\Omega^2. \qquad (A.16)$$

Here, $d\Omega^2 = d\theta^2 + \sin^2\theta d\chi^2$ is the usual differential surface element, and the parameter

$$R = 2GM$$

is written in terms of the mass of the central object, $M$, and Newton's gravitational constant $G$; the parameter $R$ is known as the *Schwarzschild radius* (or sometimes the 'gravitational radius'). Notice that, unlike Eq. (A.15) and the metrics before it, the spatial sector here is not flat. Also, and again unlike the previous expressions, this is an exact solution to an idealised problem and not a low-energy, low-mass or low-speed approximation.

The first thing to notice about this metric is that it is compatible with the weak-field solution Eq. (A.15) when $R \ll r$; that is, when the mass is small, or when an observer is far away from it. The above value of the constant $R$ is determined by demanding that the solution matches Newton's law of gravitation in this limit.

That also implies that the metric is 'asymptotically flat', in the sense that, at large $r$, it reduces to the Minkowski metric. That means that, as long as $R/r$ is reasonably small, it is natural to interpret the coordinates $r$ and $t$ as being approximately the same as the corresponding coordinates in Minkowski space.

In physical units, the parameter $R$ is $R = 2GM/c^2$. For the Sun, the so-called 'gravitational parameter'[10] $GM_\odot = 1.327 \times 10^{20}\,\mathrm{m^3\,s^2}$, giving a value for the Schwarzschild radius of the Sun as

$$R_\odot = 2.95\,\mathrm{km}.$$

As you can see, the ratio $R_\odot/r$ is therefore an extremely small number everywhere in the solar system, so that the Schwarzschild solution is an excellent approximation to the spacetime in our vicinity. It is this solution that is used for all practical relativistic corrections to newtonian physics.

---

[10] The combination $GM_\odot$ is what governs the motions of objects in the solar system, and thus planetary observations can determine the value of this to great accuracy, to one part in $10^{10}$. The mass of the Sun in kilogrammes is obtained by dividing this number by $G$, but the gravitational constant is known only to around one part in $10^5$, from terrestrial measurements; so the mass of the Sun in kilogrammes has the same uncertainty. This means that the mass of the Sun is more precisely known in metres than it is in kilogrammes. See the IAU's list of 'current best estimates' at https://iau-a3.gitlab.io/NSFA/NSFA_cbe.html. For reference, the gravitational radius of the Earth is $R_\oplus = 8.86 \times 10^{-3}\,\mathrm{m}$.

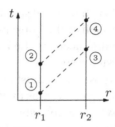

**Figure A.10** A pair of outward-going light-rays.

### A.4.3.1 Gravitational Redshift

Consider the Minkowski diagram in Figure A.10, which represents a pair of events ① and ② happening at a single location, at radius $r = r_1$. Each of these events produces a flash of light which travels outwards along a null geodesic to the events ③ and ④, which are also co-located, at radius $r = r_2$. What are the path-lengths between these various events?

The two events ① and ② happen at the same location, as do ③ and ④, so $dr = 0$, and the interval between them in the figure is the proper time separating them (that is, the time showing on a clock which is present at both events by virtue of staying still at the radius $r_1$), and this is obtainable directly from the metric Eq. (A.16):

$$\begin{aligned}
\tau_{12}^2 &= (1 - R/r_1)\Delta t_{12}^2 \\
\tau_{34}^2 &= (1 - R/r_2)\Delta t_{34}^2.
\end{aligned} \tag{A.17}$$

Since the metric is not changing in time, the pair of events ② and ④ must stand in the same relation to each other as the pair of events ① and ③, and thus the time coordinate at ④ must bear the same relationship to that at ② as the coordinate at ③ does to ①; that is, we must have $\Delta t_{34} = \Delta t_{12}$. But this means that $\tau_{12}$ is slightly *less* than $\tau_{34}$. That is, more time elapses between events ③ and ④, on the watch of an observer at $r = r_2$, than elapses between ① and ② on the watch of an observer at $r = r_1$; if the time interval $\tau_{12}$ is the period of an EM wave, then this implies that the wave will change its frequency between $r_1$ and $r_2$. Rewriting Eq. (A.17), we find

$$\frac{\tau_{12}}{\tau_{34}} = \left(\frac{1 - R/r_1}{1 - R/r_2}\right)^{-1/2}. \tag{A.18}$$

This is gravitational redshift or, if you prefer, gravitational time dilation, and is the quantitative equivalent of the argument of Section A.1.3. It must be taken into account when making precise timing observations of events both on Earth and in space. When national standards bodies compare their atomic clocks, in order to establish the consensus International Atomic Time (TAI), they must take into account the altitude, on Earth, of their observatories. And when systems such as GPS or Galileo broadcast their time signal for navigation purposes, they must include both a correction for the altitude of the spacecraft (orbital radius of 26 600 km, for GPS), which causes spacecraft time to advance with respect to the Earth at 45.7 μs per day, and a correction from SR due to the orbital velocity, which causes the spacecraft to lose 7.2 μs per day, for a total GPS advance of 38.5 μs per day.

### A.4.3.2  Geodesics in Schwarzschild Spacetime: Orbits

The Schwarzschild solution includes some geodesics which spiral around the $t$-axis in Figure A.12 – that is, they are 'orbits'. It is possible to calculate what these orbits are (another calculation we must skip here), and we can discover that, in terms of their path in the $(r, \theta, \chi)$ part of spacetime, they are nearly ellipses. They deviate from elliptical orbits to the extent that they *precess*, with the principal axis of the ellipse slowly moving around the central mass, at a speed of $\Delta\chi$ per orbit, where

$$\Delta\chi = \frac{6\pi M}{a(1 - e^2)}, \tag{A.19}$$

where $a$ is the semi-major axis and $e$ the eccentricity, as usual.

The nearest planet to the Sun is Mercury. The actual orbit of Mercury is not quite a kelperian ellipse, but instead is observed to precess at 574″/century. This is almost all explicable by newtonian perturbations arising from the presence of the other planets in the solar system, and over the course of the nineteenth century most of this anomalous precession had been accounted for in detail. The process of identifying the corresponding perturbations had also been carried out for Uranus, with the anomalies in that case resolved by predicting the existence of, and then finding, Neptune. At one point it was suspected that there was an additional planet near Mercury which could explain the anomaly, but this was ruled out on other grounds, and a residual precession of 43″/century remained stubbornly unexplained.

Mercury has semi-major axis $a = 5.79 \times 10^{10}$ m $= 193$ s, and eccentricity $e = 0.2056$. Taking the Sun's mass to be $M_{\odot} = 1.48$ km $= 4.93 \times 10^{-5}$ s, we find $\Delta\chi = 5.03 \times 10^{-7}$ rad/orbit. The orbital period of Mercury is 88 days, so that converting $\Delta\chi$ for Mercury to units of arcseconds per century, we find

$$\Delta\chi = 43\overset{''}{.}0/\text{century}.$$

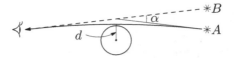

**Figure A.11** Deflection of light near a mass.

Einstein published this calculation in 1916 (Einstein 1916). It is the first of the classic tests of GR, which also include the phenomena we will look at in the next section.

### A.4.3.3 Geodesics in Schwarzschild Spacetime: Photons

We have looked at the orbits of massive particles in the Schwarzschild space-time, but massless photons also have their paths changed by the change in the spacetime (as we might expect by recalling the thought experiment of Section A.1.2). The calculations described in this section can be done exactly in the Schwarzschild spacetime or approximately in the weak-field limit of Section A.4.1.

In particular, by the same calculation process that led to Eq. (A.19), a photon which passes nearby a gravitating mass, $M$, with impact parameter $d$ (see Figure A.11) will be deflected by an angle $\alpha$, where

$$\alpha = \frac{4GM}{d}. \qquad (A.20)$$

This deflection is observable: a star which is known to be at position $A$ in Figure A.11 will be observed to appear in a telescope at position $B$. Such an observation, however, is far from easy. Taking the mass and impact parameter in Eq. (A.20) to be the mass and radius of the Sun, we find a deflection angle, for a photon which grazes the limb of the Sun, of $1\rlap{.}''75$ (if you calculate this for yourself, remember that we are using units where $c = 1$). One way of observing this is to measure the position of a star which appears near the Sun during a solar eclipse (allowing us to see the star, with the Sun's light cut off), and compare it to the same star's position when the Sun is elsewhere in the sky.

Such an experiment was performed in 1919, only a few years after the 1915 publication of GR (and incidentally after Einstein's early paper (1911), where he first predicted that there would be a deflection of light due to gravity, deriving a (mistaken) figure of half of that in Eq. (A.20)). Arthur Eddington, working with Frank Dyson (respectively directors of the Cambridge and Greenwich observatories), organised expeditions to Principe, off Western Africa, and to northern Brazil, to observe the 1919 eclipse and measure

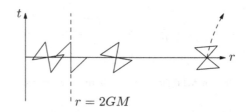

**Figure A.12** Light cones in a Schwarzschild spacetime. We also see a world-line of a particle escaping outwards.

the deflection. They found a deflection which matched the predicted one, which was widely taken as a robust confirmation of the validity of Einstein's theory. There is a little more to say about this in Section B.3.

### A.4.3.4 Black Holes

A final thing to notice about this metric is that, at the radius $r = 2GM$, the coefficient of $dr^2$ is divergent – there appears to be a singularity here. Initially, it was thought that something extraordinary happened to spacetime at that point, but further study revealed that it is merely the coordinate system that goes wrong there, in much the same way that the coordinates on the Mercator map 'go wrong' at the poles. In fact the spacetime, as observed locally, is perfectly regular at this radius, and an observer falling past this point would not notice anything amiss (that the spacetime is well behaved at $r = R$ is far from obvious, and it took several years, after the Schwarzschild solution was published, before this good behaviour was confirmed).

Is there anything special about this radius? In Figure A.12 we can see a representation of the light cones, or null cones, of an observer at various radii (see Section 4.7). These represent the range of possible future worldlines for the observer, and we can see that the effect of the Schwarzschild spacetime is to tilt these light cones by an increasing amount as we move in towards the mass. For $r > 2GM$, you can see it is possible to have a worldline which heads towards larger $r$, if it moves quickly enough. At $r = 2GM$, however, this becomes impossible: even a particle travelling at the speed of light, and thus along the diagonal of the light cone, can only remain at $r = 2GM$; and further inwards of this, at $r < 2GM$, the only possible worldlines are inward-pointing.

The Schwarzschild radius $R = 2GM$ is the radius of the *event horizon*. This close in to the centre, even a photon cannot travel quickly enough to get onto an outward-directed trajectory: even light cannot escape at this point. The Schwarzschild solution – or more precisely the spacetime when the central mass is physically compact enough that it is entirely contained within the radius $2GM$ – is the spacetime of a *black hole*.

### A.4.4 Gravitational Waves

Up to this point, we have discussed metrics which are static, meaning (slightly loosely) that they are constant in time, and which arise from the presence of a central mass. The next important solution to look at is not static, and has no source mass: it is an oscillatory solution, that of *gravitational waves*. Yet again, we cannot usefully go into any mathematical details, and can only describe the main results.

If we consider Einstein's equation, Eq. (A.10), in the absence of a source term (thus $T = 0$), and once again in the weak-field limit of Eq. (A.15), then it is possible to cast the equation in the form of a wave equation, $\nabla^2 X = 0$, where the $X$ is a tensor rather than a scalar, and the $\nabla^2$ is an appropriate laplacian operator (this is where the mathematical complications come in[11]). This equation admits of solutions which describe curvature in spacetime in the *absence* of mass, where the curvature of one part of spacetime causes curvature in a neighbouring part, in a way broadly similar to the way an electric or magnetic field will induce a magnetic or electric field in a neighbouring part of space.

Thus these gravitational waves *propagate* through spacetime at the speed of light, and carry energy. In particular, they carry energy *away* from a source consisting of one or more accelerating masses; the masses in question are most typically a pair of compact masses, such as neutron stars or black holes, orbiting each other. It is possible to predict the amount of energy being carried away from such a system by gravitational waves, as a function of the masses and orbital parameters of the objects.

The binary pulsar PSR 1913+16 (also known as the 'Hulse–Taylor binary') is a pair of neutron stars orbiting each other. The arrival times of the signal from the pulsar have been measured to high accuracy over the decades since the binary was discovered, and the orbital period has been found to be increasing – the binary is slowing down. The behaviour of the binary pulsar, and in particular the rate of spindown, is consistent with GR to high accuracy, and this, in the absence of any other mechanism for energy loss, is taken as an early indirect test of the existence of gravitational waves.

What do gravitational waves look like?

---

[11] To be honest, this is where they come in and kick over the furniture and scare the cat.

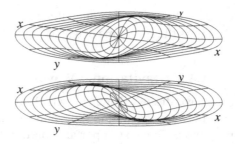

**Figure A.13** A sketch of a gravitational wave.

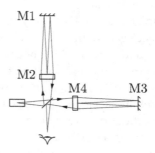

**Figure A.14** An interferometer.

In Figure A.13 we can see a sketch of the analogue of a gravitational wave in a two-dimensional space, in two different phases of its oscillation. If you imagine measuring the diameters of the top 'circle' in the $x$- and $y$-directions by pacing from one end to the other, you will see that they will be different from the $x$- and $y$-diameters of the bottom 'circle'. If we placed a set of test particles around the edge of this circle, then the distances between them would change over the period of a gravitational wave, even though they did not move, and specifically *did not accelerate*. If the test particles were mirrors, at either end of the cross drawn at the bottom of the figure, and we shone a laser between them, we would expect different times of flight between the mirrors at different phases of the passing gravitational wave.

We have just described an interferometer, of the type which is used in the LIGO and Virgo collaborations to attempt to directly detect gravitational waves. In the interferometer sketched in Figure A.14, we see a laser shining into a beam-splitter, which sends the beam down both the arms of the apparatus, where it is reflected back and forth between the four mirrors M1 to M4, before being recombined at the detector, at the bottom. All four mirrors are carefully suspended, so that they do not vibrate from any local, terrestrial, cause. If a gravitational wave passes, then one or both of the M1–M2 or M3–M4 distances will change, in the way we have seen above, without any of the mirrors accelerating. Thus the path length of one or other arm will change, causing a detectable change in the interference pattern at the detector. The two LIGO detectors, in Louisiana and Washington State in the USA, each have arms 4 km long and send the light back and forth hundreds of times. It is a fine irony that this apparatus is broadly the same design as the Michelson–Morley experiment which made its appearance at the beginning of the history of SR.

The measurement is a little tricky, however, since the anticipated change in the 4 km arm length, for a plausibly detectable gravitational wave source, is of the order of $10^{-18}$ m. After decades of effort, on improving lasers, mirrors, mirror suspensions, and data analysis techniques, this measurement was finally made, and jointly announced by the LIGO and Virgo collaborations in 2015. In doing so, the collaborations not only directly confirmed the existence of gravitational waves, and thus further corroborated General Relativity, but also opened up a new non-electromagnetic messenger (alongside neutrinos and cosmic rays) for the observation of the universe.

### A.4.5 Cosmology

In the last few sections, we have looked at the spacetime around a small mass, around a large mass, and waves in a spacetime produced by an accelerating mass. The next step: the universe!

To find a metric for the universe as a whole, we start with the observation that, on the largest scales, the universe appears to be homogeneous and isotropic – that is, it is the same at all spatial points, and in all spatial directions. This is known as the *copernican principle*. Apart from encouraging very proper modesty about our place in the universe, this permits a considerable mathematical simplification. It is possible to write down, from purely mathematical considerations, the most general metric which has these two properties, namely the Friedman–Lemaître–Robertson–Walker metric.[12] In the most convenient choice of coordinates, it looks like

$$ds^2 = dt^2 - a^2(t)\left[\frac{dr^2}{1 - \kappa r^2} + r^2 d\Omega^2\right], \tag{A.21}$$

where $d\Omega^2$ is the angular area element, as before, $r$ is a radial coordinate,

---

[12] Usually referred to as 'FLRW', or by whatever subset and ordering of the four independent discoverers' names is demanded by your national prejudices.

scaled so that the parameter $\kappa$ has one of the values $\{-1, 0, +1\}$, and $a(t)$ is a dimensionless scaling factor which depends only on the time coordinate.

The effect of the scaling factor $a(t)$ is, in effect, to change the size of the universe. It turns out that, when we apply Einstein's equation, Eq. (A.10), to this metric and a reasonable energy-momentum tensor, we deduce $\ddot{a} \neq 0$, and thus a universe which is *changing in size*.

That Einstein's equations admit of a non-stationary solution – an expanding universe – was initially thought to be a physically implausible defect in the theory. To amend this, Einstein observed that the argument leading up to Eq. (A.10) can be slightly extended to allow an additional term on the left-hand side, $\Lambda g$, including a scaling factor now known as the *cosmological constant*; this modified version of Einstein's equation does admit of a stationary solution, albeit an unstable one. The discovery of the Hubble–Lemaître expansion of the universe meant that the need for this fell away, but the last couple of decades have seen the term reintroduced, responding to otherwise-inexplicable observational evidence, and corresponding to a further exotic source of energy-momentum in the universe, not identifiable as either matter or radiation.

Moving from maths to physics, the next step is to consider various dispositions of energy-momentum, and then use Eq. (A.10) to constrain the form of $a(t)$, and thus its first and second time derivatives $\dot{a}$ and $\ddot{a}$, with different solutions depending on the value of $\kappa$. At that point, we cross over into the study of cosmology.

One extreme case of this metric is where $a(t) = 0$. The solutions to the FLRW metric generally include an in-principle initial state where this is so, and they have a singularity at this point. This is the *big bang*. At this point, the curvature of the metric becomes infinite, which implies an infinite energy density, which is enough to disrupt some of the mathematical foundations of GR. At this point, quantum effects become important, obliging us to attempt to develop a theory of quantum gravity. That is the next challenge.

# Appendix B
## Relativity's Contact with Experimental Fact

For, if I had believed that we could ignore these eight minutes, I would have patched up my hypothesis accordingly. But since it was not permissible to ignore them, those eight minutes point the road to a complete reformation of astronomy: they have become the building material for a large part of this work....

*Johannes Kepler, on the discrepancy of* 8′ *of arc between the predicted and actual positions of Mars, in* Astronomia Nova, *II, Cap.19, quoted in Arthur Koestler,* The Sleepwalkers *(1959)*

Special Relativity is a physical theory – a theory of how our universe works. It is a hugely successful theory, which has been corroborated both implicitly and explicitly by numerous high-precision measurements in the years since the theory was first proposed.

That said, SR's contact with experiment is a little odd, compared with other parts of the physical sciences. It is foundational to physics, and its validity essentially unquestioned, to the extent that there is paradoxically rather little to test: any deviation between SR and experiment would require very large areas of physics to be rethought. This means that what current efforts there are, are not asking routine questions of 'is this theory correct?'. Instead, they are looking for any discrepancies at all since, however small or subtle such differences might turn out to be, they would be ground-shifting.

This means that when we describe 'experimental support for SR', we are almost always doing so with pedagogical or historical intent, rather than as part of an open claim about validity. That pedagogical intent can be to deepen our understanding of the physics, but in this appendix I want instead to discuss the way in which the link with experiment throws up questions of what it means to test a theory. For me, the interesting thing about the Michelson–Morley test (mentioned in Section 2.1.2), and to a greater extent about the Dyson–Eddington eclipse test (Section A.4.3.3 and below), is that they show the extent to which science is a social as well as a logical process: to the extent there is a 'scientific method' which demarcates science from non-science, it's a social one of asking certain types of questions which are answerable in a particular way, with these questions being asked within a community which aims and is able to arrive at provisional consensuses by a more-or-less systematic route.

'What is science?' is an interesting question, but it is not a scientific one. However, this question, along with 'what does science know?', 'how does science justify its statements?', and 'how do scientists choose between theories?' are all importantly *different* questions, however similar they may sound at first. The last of these questions is the most interesting one, because it is perhaps the one least often asked. I don't plan to offer any confident opinions about these questions – they are too large for that – but I hope to illustrate that they are distinct questions, and to persuade you that the last one is one that should be asked more often.

A very conventional starting point, when talking about science, is to observe (this is Karl Popper's formulation) that science proceeds by 'conjecture and refutation', meaning that the 'scientific method', such as it is, consists of theories being proposed, and then those theories being either refuted by experiment, or not refuted yet. Though this badly oversimplifies the practice of science – there are epistemological, methodological and even sociological complications that have to be taken into account – it does capture the essential asymmetry of scientific logic: experiments can only prove theories wrong, but never really prove them right, except from exhaustion. So relativity's contact with experiment consists of searching for an experimental failure.

## B.1  Special Relativity

As we have seen earlier in this text, relativity rests on only two physical statements, which we can rephrase here as:

1. all inertial frames are equivalent, and there is no preferred frame;
2. – there is an upper limit to speeds in nature;
   – or, the speed of light is the same for all observers;
   – or, the elapsed time measured in a moving frame is $t' = t/\gamma$;
   – or, the interval $s^2$ is invariant between inertial frames

(see the discussion in Section 5.8.1). Tests of SR consist of attempts to devise an experiment which lets us observe one or other of these statements, or their necessary consequences, to be false. We only have to find one failure, for physics to be turned on its head.

The first link is electromagnetism (EM). As I described in Section 5.8.2, Maxwell's equations are not invariant under the galilean transformation; the Lorentz transformation was developed, before Einstein's relativity, in order to establish the transformation that *did* leave Maxwell's equations invariant; and it is possible to imagine an alternative history in which SR was developed by extrapolating from this invariance of electromagnetism to the rest of physics. From a point of view which embraces SR as foundational, however, the experimental success of Maxwell's equations, and the fact that they are Lorentz invariant, is not merely a trigger for SR, but a powerful corroboration of it. Einstein demands that any plausible physical theory must be Lorentz invariant, and – look! – we already have a successful theory which illustrates that.

Indeed, it is possible to start with the non-relativistic Lorentz force law $\mathbf{f} = q(\mathbf{E} + \mathbf{v} \times \mathbf{B})$, and Coulomb's law for a stationary charge, $\nabla \cdot \mathbf{E} = 4\pi\rho$, add the demand of Lorentz invariance, and deduce Maxwell's equations as a consequence.[1] In a way, it's Maxwell's equations that are the first experimental test of SR.

There is a difference of interpretation, of course. Starting from SR, it is (fairly) natural to develop a view where space and time are facets of a single structure, rather than rigidly demarcating time from space, in the way that (what is now called) the newtonian picture does. This is quite a change of perspective, and a disorienting one, and it was natural to be sceptical. Lorentz and his colleagues did not think of Eq. (5.16) in terms of spacetime rotations, of course, but it meant they were obliged to come up with mechanisms which would contract measuring rods in the direction of motion, and slow down clocks. FitzGerald's and Lorentz's suggestion,[2] that the aether somehow compressed measuring rods in the direction of travel, can be regarded as an adequate temporary hypothesis, but one which started running out of steam when experiments failed to demonstrate the aether behaving as consequently required.

The 1887 Michelson–Morley experiment used an optical arrangement very much like the LIGO and Virgo experiments of Figure A.14, but with arms a few metres long. Michelson and Morley expected to demonstrate

---

[1] This is illustrated in Rindler (2006, §7.3). It is illuminating, but the maths is a little beyond the level of this text.

[2] See FitzGerald (1889) and Lorentz (1895); Lorentz reports that he had first suggested this idea in 1892, and also notes that Oliver Lodge referred to the idea in 1893.

that the effective light path changed length when the arms were moving or not moving with respect to the aether, resulting in observable interference effects. If they had observed this, they would have *refuted* the statement that all inertial frames are equivalent – they would have shown that the aether frame had a special status, and that Maxwell's equations only work properly in that frame. Famously, they did not see this, and nor did the various people who attempted to re-do the experiment in one or other variant, encouraged at various times by both Lorentz and Einstein. Attempts were made with instruments of different materials, at altitude, housed in buildings with thick or thin walls, at different seasons of the year, and using sunlight or starlight instead of artificial light, in case any of these made a difference. In no case, as far as I am aware, was there any principled reason why such a difference would be expected; for all that they were experimenting with light, these observers were exploring in darkness. As late as 1933, Dayton Miller claimed to have detected an absolute motion with respect to the aether, which was smaller than expected but still inconsistent with zero, requiring (he asserted) adjustments to either the aether theory or the FitzGerald–Lorentz contraction, but still in contradiction to the precisely-null result required by relativity theory (Miller 1933). Miller's one-time research assistant reanalysed the data in 1955, however, and concluded that, despite all efforts, the measurements were still consistent with a null result (Shankland et al. 1955), and thus that there was still no refutation.[3]

I've described this history at a little length, partly because it's a famous and important experiment, but also because it is illuminating about the way that the status of a measurement changes within the scientific community. As usefully discussed in Collins & Pinch (2012, ch. 2), Michelson and Morley's null result was initially understood as an anomaly for aether theory; later, it was re-perceived as consistent with relativity, and a confirmation of it; finally, Miller's claimed non-null result changed, and from being seen as an anomaly in relativity theory, it itself became an experimental anomaly, to be explained away by Shankland et al. as a systematic observational error.

---

[3] Both Miller's and Shankland et al.'s papers provide fascinating and detailed accounts of the history of this measurement. Reading these papers one is repeatedly reminded of the technical challenges of highly sensitive equipment, challenges which were and are also faced by the gravitational wave communities, who also experienced frustrations and disruptions from distant traffic, microseismic events, weather, and gruelling observational campaigns. Rather desperately in retrospect, Miller ends his paper with a list of other observations which seem to hint at an absolute Earth motion, the last one of which, ironically, is Karl G. Jansky's report of 'a peculiar hissing sound in short wave radio reception, which comes from a definite cosmic direction'.

The Michelson–Morley experiment is at heart an EM experiment and, although this is not the way it is usually framed, it shows that EM really is fully consistent with the Lorentz transformation.

After electromagnetism, the next and possibly most powerful implicit test of relativity is its tight integration with other fundamental areas of physics. If Special Relativity were not correct, then General Relativity, which builds precisely on top of it, would be incorrect, too. However, SR is also the foundation on which relativistic quantum mechanics (RQM) is built, which is in turn the foundation, alongside classical field theory, for the gauge field theories of the Standard Model of particle physics. RQM is quantum mechanics rebooted on the assumption that it is taking place in Minkowski space. If this assumption were not correct, then none of contemporary particle physics could be observed, particle accelerators would not work as they do, and – put most crudely – the engineers who built CERN would have different mechanical and electrical problems to solve.

There is a further well-known implicit test in the satellite navigation systems such as GPS, GLONASS, and Galileo. These work by each of the network of satellites transmitting its precise position and time at regular intervals; receivers pick up these signals from four or more satellites, and can deduce their current position by comparing the multiple reports (Ashby 2003). As we saw near Eq. (A.18), the clocks on board such a satellite advance, relative to clocks on the ground, at about 38 µs per day. That immediately translates to a position error, if uncorrected, of 38 light-microseconds (in units where $c = 1$), or 11 km per day in physical units. That the receiver in your phone is more accurate than this is not because your phone is doing any relativity calculations, but because the satellites are designed to report their current time with this correction factored in. In addition, the satellite clocks are running in an inertial frame, but transmitting their reports to, and receiving calibration updates from, a rotating Earth. For this to work, the system as a whole has to take account of the Sagnac effect, which changes the phase of a light signal which propagates in a rotating reference frame.[4] These satellites, like the design of CERN, are not designed as 'tests' of SR; but as a fact of engineering none of these systems would work if the spacecraft clocks, and their relationships with clocks on the Earth's surface, did not behave as SR and GR predict.

---

[4] The Sagnac effect is also used for practical navigation in a class of high-precision gyroscopes called 'ring-laser gyroscopes'. The Sagnac experiment was devised by Georges Sagnac in order to demonstrate the presence of the aether; it appeared to, at the time, but turns out to corroborate relativity on closer examination. The Sagnac effect is discussed in Ashby (2003), and there is a history in Pascoli (2017).

In the presence of these powerful implicit tests of SR, I don't feel there's really a great deal extra that's useful to say about the various explicit ones. There is an extensive annotated bibliography of experimental tests by Roberts & Schleif (2007) and a summary in Will (2014, §2.1.2): it seems sufficient to say that SR has been tested with great ingenuity, at length, and to great precision, and it has not failed any test so far. At the end of Roberts & Schleif's bibliography is a section on papers which claim but do not sustain results inconsistent with SR, which is worth reading as a series of (unwitting) demonstrations of how hard experimental work is, in this area.

Although the underlying theory is not in doubt, there nonetheless *are* significant current efforts to make measurements which look for deviations from SR's predictions, using observations which stretch from sub-nuclear physics to astrophysics. Various exotic physical theories – and there is no current shortage of these, mostly to do with quantum gravity, theories of the quantum mechanical structure of spacetime at the microscopic level, or exotic cosmologies, but also including theories of particle physics at very high energies – include predictions of some violation of Lorentz invariance. That is, a proposed theory includes some preferred frame (in the sense that the aether was a 'preferred frame', of absolute rest) or, what is much the same thing, that equations of motion or measured physical properties would have different values in different inertial frames. The goal here is not to 'prove Einstein wrong', but instead to discover some way in which the axioms of SR are inadequate or incomplete, and so to uncover new physics beyond the Standard Model. Finding a preferred frame would contradict the first postulate, above, turning it from a fundamental organisational principle of our universe into a special case or low-energy approximation. No matter how special the special case, or how precise the approximation, if any deviation from perfect Lorentz invariance were found, that would be of major significance, and provoke sudden huge interest in any theory which predicted it; if no deviation is found, then that usefully rules out those theories which propose it or depend upon it; for a review, see Mattingly (2005). All of these experiments are technically challenging. None so far has found anything.

## B.2 General Relativity: Classical and Post-classical Tests

Unlike Special Relativity, General Relativity is not so embedded into the fabric of physics, that doubting it verges on the irrational. Although the two subjects are fundamentally linked in their concerns and approach, and although GR builds directly on top of SR – remember that GR is the attempt to repeat the success of SR outside of the special case of no gravity, and that Statement A.9 has SR acting as the 'bridge' into GR – SR sits at the centre of physics while GR stands at least a little way closer to the border. There is very little about SR that is in any sense optional: the first axiom is a deeply fundamental principle which it seems reckless to reject, and the second is a binary choice between there being an upper speed limit in nature, or not. If SR isn't true, very little makes sense.

In contrast, there is significantly more that is contingent in GR. The version of the EP in Statement A.9 is a physical statement, which *appears* to be extensively confirmed by experience, but which *might* be wrong – perhaps there really are curvature-coupling terms in some exotic circumstances. Einstein added a 'cosmological constant' to the field equations, as we saw in Section A.4.5, so there is already some choice in the matter of the field equations to be solved, and it is possible in principle that our Eq. (A.10) is incomplete. The argument leading up to Eq. (A.10) stresses the extent to which Einstein's equation is a (well-motivated) guess: experiments corroborate the physical validity of this guess, but while a failure to confirm some detailed prediction of GR would be an event of extraordinary importance, it would not be the fundamental earthquake represented by any failure to confirm SR.

The three so-called 'classical tests' of GR are:

- the advance in the perihelion of Mercury, beyond that predicted by newtonian mechanics, as we saw in Section A.4.3.2;
- the deflection of light by the Sun, as we saw in Section A.4.3.3, and as observed by Eddington and Dyson in 1919; and
- the red-shifting of light, which was finally directly confirmed in the Pound–Rebka experiments of 1959–64, as we described in Section A.1.3.

These tests' 'classic' status derives in part from Einstein's description of them in the final section of his 1916 paper, but also because the successful tests are famous, and between them seem generally regarded as demonstrating GR to be correct.

It is remarkable how quickly and completely the scientific community arrived at the consensus that Eddington and Dyson's eclipse measurements had confirmed the correctness of GR. Though some scientists retained a principled scepticism, there is little sign of serious scientific disagreement after the compatible results from the (much less famous) 1922 eclipse; and the similar Yerkes Observatory observations of the 1952 eclipse, and the Texas observations of the 1973 one, were reported as routine confirmations of a well-established result. This is all the more remarkable because the eclipse measurement is very challenging, and remained at the limit of what was measurable whenever it was repeated – the 1952 and 1973 observations reported similar errors (around 10%) to the 1919 ones. The light-bending measurement did not become routine until the emergence of radio and pulsar astronomy, which quickly achieved sub-arcsecond position resolution: since the Sun isn't dazzlingly bright in radio, there is no need to wait for an eclipse, so that measurement of the solar deflection becomes straightforward. The deflection of electromagnetic radiation by the Sun, and indeed the general effects of the line of sight passing through the Sun's Schwarzschild spacetime, are now merely standard observational corrections for radio astronomers.

Confirmation of gravitational redshift took longer. In comparison with the fame of the 1919 eclipse measurements, it is less often mentioned that gravitational redshift should produce a measurable shift in the absorption lines in the solar spectrum, and that attempts to observe these, in the 1910s, produced contradictory results. The theoretical prediction, on the part of various authors including Eddington, was unclear (surprisingly so, given how straightforward the argument now seems in retrospect, in Section A.1.3), but did agree that there should be a redshift (and this was the consensus after 1921); and an open question at the time was whether quantum phenomena (i.e., photons) would experience relativistic effects in the same way that classical electromagnetic fields do (again, this seems fairly obvious from our point of view in the twenty-first century). By 1919 multiple observers had attempted to detect the redshift in the Fraunhofer lines, without clear success, and until the mid-1920s it was possible to enlist these observations on both sides of the argument. Although the consensus from then appears to have moved towards corroboration (more so after the 1919 observations), it was only in the 1960s that the redshift was measured convincingly. See Earman & Glymour (1980a) and Hentschel (1996) for the history, and Will (2014, §2.1.3) for discussion and further references.

It was also in the 1960s that it became technically feasible for Pound, Rebka and co-workers to directly measure the gravitational redshift in a terrestrial laboratory, as mentioned above. Other GR tests, from a further detection of redshift by Gravity Probe A in 1976 to the detection of gravitational waves in 2015, are demonstrations of huge experimental skill, at the edge of what is measurable, but in no case has the outcome been seriously in doubt. All scientific statements are in principle permanently provisional, but a consensus emerges, in a particular area, when it becomes fruitlessly difficult to generate or sustain plausible alternatives to the core theories. That GR is 'correct', in this sense, has been a settled fact since about 1919.

So is that the end of GR's confrontation with experiment? No, it's not, but the nature of the confrontation has changed.

The 'classical' tests represented the first phase of the theory's confrontation with reality, addressing the question 'is this theory correct?'. As we have seen, this phase effectively finished in 1919 when, broadly speaking, the statement 'GR is correct' became conventional knowledge, and the subject became something one could consider teaching to undergraduates. Then there was a 'mopping up' phase, during which the remainder of GR's predictions were elaborated and then tested. There was little expectation that these tests would show a failure in the theory, but they were interesting and valuable as technical challenges which would tax and advance other sub-disciplines within physics, always with the off-chance of uncovering a thrillingly unexpected anomaly. This phase might be said to have concluded with the detection of gravitational waves, and that sub-discipline's transition from a test of GR, to another messenger for astronomy. This broadly fits in with the historically motivated account of the social dynamic of science famously described by Thomas Kuhn (1996), which sees it work round a cycle of 'revolution', where a new idea changes the ground rules, to 'normal science', a technical puzzle-solving phase where details are worked out and anomalies resolved, then to a phase where a residue of unresolved anomalies builds up to the point where it creates a 'crisis', which is finally resolved by a new conceptual revolution, and the beginning of another cycle.

So where are the 'anomalies' with GR? General Relativity is now so firmly established as a foundational theory of our universe that *any* deviation between theory and observation would be a major scientific event (if not quite the earthquake that would result from a negative result in SR). This search for anomalies is the new 'normal science' for GR, and is represented by a systematic parameterisation of the various ways that a theory of gravity might plausibly differ from Einstein's.[5] Thus the 'Parameterised Post-Newtonian'

---

[5] A similar process is currently taking place in experimental particle physics, where 'normal science' is the increasingly desperate attempt to find some measurements which are demonstrably at odds with the Standard Model of particle physics.

formulation of GR is a theoretically motivated distillation of the question 'in what ways could GR be incorrect?', which is concrete enough to allow critical measurement. So far, no deviations have been found, but the attempts to find them have been summarised at exhaustive length in Will (2014).

## B.3  A Closer Look at the 1919 Eclipse Observations

It is worthwhile spending a little time looking again at the swiftness and completeness of the community's acceptance of Eddington and Dyson's observations and conclusion. This is because it leads us to interesting questions about how science works in practice, rather than in widely accepted myth.

Eddington's and Dyson's observations of the solar eclipse of 1919 May 29 were technically challenging, and although the conclusions are solid, the observational analysis is not beyond comment. Indeed, this set of observations has become something of a case-study in the subtleties of 'theory choice' in science – how is it that a particular, potentially disputable, observation is or is not taken to be conclusive by the scientific community? That the case-study has gone on to achieve that most unwelcome type of fame, notoriety, is separately interesting.

The joint expedition was planned starting in 1917, as a collaboration between Frank Dyson, who was both the Astronomer Royal and the director of the Royal Greenwich Observatory (RGO), and Arthur Eddington, the director of the Cambridge University observatory. The expedition was planned during, and undertaken just after, the First World War, and wartime logistics meant that the equipment was not everything the observers could hope for. It is also at least interesting that Eddington, who was of conscriptable age, but who, as a devout Quaker, had made it clear he was prepared to object to military service on conscientious grounds (an unpopular position which could have resulted in him being detained), led the Principe expedition as an explicit part of a negotiated scientific deferment of his conscription (see Earman & Glymour (1980b), quoting Chandrasekhar).

The observational goal was to measure a deflection in stellar positions of around one arcsecond, in the field (literally) rather than at an observatory, and in the few minutes of totality available during an eclipse. The observations were framed to distinguish between three possibilities, namely no deflection, the $1''\!.75$ 'einsteinian' prediction of GR, and a so-called 'newtonian' deflection of half of that, $0''\!.87$. The full deflection is the prediction of Eq. (A.20), based on Einstein's field equations and a non-euclidean spatial sector; whereas the 'half-einstein' prediction arises from an argument related to that of Section A.4.1, in the sense that it depends only on the Equivalence Principle (this half value was the one that Einstein predicted when he first discussed the deflection in 1911). In his 1918 'Report' on relativity, Eddington discusses (1920, §32) the newtonian deflection only as a pedagogical contrast to the einsteinian one, so it seems likely that the latter was the value he expected to observe on the expedition; but the derivation of the result was a lot less clear in Eddington's 1918 work than it is now, and both non-zero possibilities were clearly scientifically realistic outcomes. There is a detailed description of the history of the theoretical work in Earman & Glymour (1980b, §1).

Measurements were made by Dyson's RGO colleagues at Sobral in Brazil, and by Eddington and assistants at Principe, off the coast of West Africa; Dyson was responsible for the data reduction for the Brazil observations, and Eddington did the data reduction for Principe. There were three telescopes taken to the two sites: an astrographic (survey) instrument in both places, each equipped with a clockwork coelostat, plus, in Sobral, a 4-inch instrument with a narrower field of view.

The results were reported briefly to a joint Royal Society and Royal Astronomical Society meeting on 1919 November 6.[6] Despite being originally intended as a backup, the 4-inch instrument produced the cleanest results, giving a deflection of $1''.98 \pm 0''.18$.[7] Both of the astrographic observations had problems, partly to do with cloud cover and focus problems, and partly because of the technical need for a more elaborate data reduction process; and Eddington and Dyson reported (1920), deflections of $1''.61 \pm 0''.30$ for Principe and $0''.93$ (no error quoted) for Sobral. Dyson concluded that, although the Sobral astrographic result was closest to the 'newtonian' figure, the result was too unreliable, after analysis, to give much information, and it was effectively discarded.[8] Eddington's analysis led him to the large error

---

[6] The joint meeting is briefly summarised in Thomson (1919), which includes a record of the discussion afterwards. This was accompanied by a much more detailed account (Dyson et al. 1920), which was 'read' to the Royal Society on the same date in 1919. You can find fuller accounts of the expedition and experimental analysis in Earman & Glymour (1980b), and in a brief but lucid account in Kennefick (2009), which is related to a more detailed version, which also talks about the 'myths' surrounding the observation, in Kennefick (2012).

[7] It was usual at the time to quote errors in terms of 'probable error' rather than standard deviations: the probable error is half the interquartile range, and for a normal distribution is a little more than 2/3 of the standard deviation. Here and below I have quoted the errors rescaled to standard deviations.

[8] The Sobral plates were remeasured at the RGO in 1979 (Harvey 1979) with the result that the error on the 4-inch observations was tightened, and the result from the (Sobral)

quoted here, but he argued that there were no systematic errors which would otherwise undermine this result (indeed that the cloud cover at Principe had accidentally but usefully avoided the overexposures which had been part of the problem at Sobral). Specifically, he stated that 'the accuracy seems sufficient to give a fairly trustworthy confirmation of Einstein's theory, and to render the half-deflection at least very improbable', but that this Principe result 'has much less weight than [the 4-inch one].' The logic of the report, in its conclusion, is that the 4-inch observation is the main result, inconsistent with the 'newtonian' prediction, that this is weakly corroborated by the Principe data, and that the Sobral astrographic observation was too unreliable to rule out either possibility.

I have described this measurement in much more detail than you will usually find in a text of this type – relativity textbooks tend to deal with the 1919 measurements in a sentence or a short paragraph – but there is little that is really unusual about it. It was a challenging measurement at the limit of what was possible; the observers constructed and published an argument that some of the data was unreliable and should be discarded, and that some other data, though difficult to analyse, should be retained; they came to a qualified but still confident conclusion about its consistency with the 'einsteinian' result; and the community promptly and overwhelmingly accepted this conclusion, and seems to have regarded the matter as largely settled. It is still regarded as a famously conclusive measurement.

After discussing the history of the measurement, Earman & Glymour (1980*b*) drew attention to a number of features of this measurement which are not uncommon in scientific experimentation, but which were particularly clearly exhibited here; their account has been much discussed since.

(i) The deflection measurements were motivated by a particular theoretical prediction, and so were designed to rule in or rule out three distinct possibilities, rather than making an unconstrained estimate of the value. In other words, this was a hypothesis-testing experiment, and not the measurement of a physical parameter.

(ii) Although they were in the service of a particular prediction, there was no widespread understanding of the details of the theory. Eddington was one of the few people able to do the calculations confidently, but even there, it is clear that Eddington's, Einstein's and others' understanding of the theory, and of how to do calculations, was not as profound as it would later become. That is, it would have been reasonable to criticise the theoretical

---

astrographic plates improved to $1''\!55 \pm 0''\!34$, both ruling out the 'newtonian' prediction and suggesting that Dyson had been pessimistic in his analysis.

work, though no-one seems to have done so in fact. It appears likely, in context, that the consistency of the measurement with the theory would have increased confidence in the validity of the derivation of the predictions from the theory, even though this would be circular in strict logical terms.

(iii) Eddington was already convinced of the correctness of GR, though he was in something of a minority here. Dyson was sceptical of it, as were others who took part in prior attempts to make this observation. Further, there was open antipathy to German science in general, in the UK in the aftermath of the war, and it seems very likely that a significant part of Eddington's personal motivation, and possibly part of Dyson's as well, was to 'put an end to wild talk of boycotting German science', as Eddington put it in his obituary (1940) for Dyson (the obituary, written in the early years of the Second World War, goes on to remark that the confirmation of a German theory 'kept alive the finest traditions of science; and the lesson is perhaps still needed in the world to-day').

(iv) Dyson and Eddington offered others access to their data, for reanalysis, but no-one seems to have thought it useful to do so: their analyses of their observations were taken at face value, on the grounds of their expertise (in the obituary, Eddington notes that Dyson 'had a reputation for soundness', and draws attention to his advances in plate reduction techniques).

(v) The difficult measurements of Fraunhofer redshift produced inconclusive results before 1919, but soon afterwards they were consistently found to corroborate it. The goal of such measurements had, it seems, changed: they had been unable to test the truth or falsity of the theory by themselves, but by taking GR to be more-or-less confirmed, observers became able to use these observations to achieve other scientific goals. If a particular measurement appeared to be inconsistent with the redshift, then that would be a prompt to re-examine it in search of a practical defect; but if it were consistent, there would be no such prompt (this is not an empty process: scientific progress sometimes comes when anomalies stubbornly remain even after such re-examination).

None of these features – prior convictions, theoretical uncertainty, personal motivations, data selection, the appeal to authority, and apparently sudden consensus – is unusual in science as it is actually practised; the history of science is full of other examples, and we have already seem some similar features in the fuller history of the Pound–Rebka measurements (note 6 on p. 152). What is unusual in this story is that so many of these features are present at once, and that the history of the case is so well known. In addition, these various features tend *not* to be ones mentioned in popular accounts of science (written by scientists as well as non-scientists), which usually tend to retail an account of science as an activity carried on with dry rigour – *this* observation leads to *that* conclusion – and with a linear inevitability from theory, to experiment, to certainty.

This is not a claim that science is somehow illogical (it is logical, and an assertion that observation A is inconsistent with, or 'falsifies', theory B will prompt critical examination of both[9]), or that it is unreliable (over time, science does seem to ratchet towards a better and better understanding of a certain category of knowledge). Instead, it is to assert that, when we look at a case such as the 1919 eclipse, the deductive logic of falsification is far from being the only thing going on. How is it that a community of scientists will choose one conclusion over another, at a particular point in time? The observational evidence is clearly a part of it – and science frames questions so that this evidence is a much bigger influence on its conclusions than is possible in other areas of human enquiry – but the other features are not absent, even if they play less and less of a role in our acceptance of a theory, as time moves on and we are able to see an observation in hindsight and in an ever-enlarging context. But if we want to look back, and try to understand how a particular confrontation between theory and experiment was actually resolved, at a particular historical moment, then we cannot use that hindsight to make sense of contemporary theory choice, which can only be examined, and puzzled over, in its own, contemporary, terms.

This case became famous partly because, after drawing attention to some of these features in their history of the case, Earman and Glymour remarked in passing, in their final paragraph, that '[t]his curious sequence of reasons might be cause enough for despair on the part of those who see in science a model of objectivity and rationality'. Some people have chosen to see this remark as an attack on the integrity of Science, or even on the personal integrity of Eddington or Dyson; this interpretation makes little sense, not least because the authors go on to remark bluntly, as the very last words in their paper, that GR 'still holds the truth about space, time and gravity'.

The case has also been of interest to those who are not primarily interested in the history, or the science as such, but instead interested in what it and similar cases tell us about the specifically social aspects of theory choice in the past and in the present – the Sociology of Scientific Knowledge.[10] This

---

[9] I use the word 'falsify' because that fits in with the popperian logic of science, but 'falsification' is a complicated business, and the more evasive phrasing of 'inconsistent with' is probably more descriptive.

[10] 'In the present' is important, because current controversies, perhaps scientific controversies which interact with crucial public policy decisions, necessarily happen without the benefit of any hindsight, even though 'how do we know?' and 'what do we do now?' may be

approach originated in the work of David Bloor in 1976 (i.e., Bloor (1991), and see Laudan (1981) and Bloor (1981)), and is well summarised in Collins & Pinch (2012). It represented, I suppose, one 'side' in the so-called 'Science Wars' of the 1990s, an unedifying episode which has a useful synthesis in Labinger & Collins (2001).

It's also worth pointing out that Earman and Glymour were not the first to talk of this case, nor the first to raise the suggestion (to the extent that they did) that there was something amiss in Dyson and Eddington's measurement. Indeed, it seems to have become something of an urban myth in gravitation physics: it was mentioned in a rather garbled aside by Everitt et al. (1979), in an article proposing an alternative test of GR, and it reached possibly its more extreme published form in Hawking's *A Brief History of Time* (1988), where he suggests, without pointing to a source, that 'the errors were as great as the effect they were trying to measure.' There are at least two interesting things about these two (historically inaccurate) accounts: one is that neither author thinks it important, from a scholarly point of view, to give a precise account of the history; secondly, neither author seems scandalised by the incident, and indeed Hawking goes on to remark that this was 'a case of knowing the result they wanted to get, not an uncommon occurrence in science'. When I have heard this story retold, in one variant or another, or retold it myself, it it has never been received with either surprise or outrage, but at most wry amusement or a facetious tut-tut-tut.

From one point of view, this doesn't matter. Science tends to be cheerfully ahistorical in practice, and most scientists would probably agree that, as a question of how we know what we know, it doesn't much *matter* just why the community arrived at one conclusion or the other, as long as whatever conclusion we have on the matter, now, can reasonably be agreed to be correct (this touches on the distinctions between the various questions on p. 178). Science gains its justifications from current consistency between theory and experiment, and when we hear 'Newton said...' or 'Galileo observed...' it's generally a pedagogical point being made, rather than a historical one, or any attempt to persuade. The history of science is much more intricate (and indeed more interesting) than you would generally learn in a science lecture, but that history is not part of the professed *logic*

scientific questions which demand immediate answers. I am writing this during the COVID-19 pandemic, when demands that politicians 'follow the science!', though reasonable, were met with some unease by the scientists in question, and when the contrasts between scientific, medical, and political theory-choice – 'how do we know?', 'what are the chances?', and 'what do we do now?' – were discussed in newspaper leading articles rather than journals of the sociology of science.

of science.

The reason I am describing this at such length is that I think that the episode illustrates some important features of the relationship between theory and experiment, in actual scientific practice. In the swift acceptance of the 1919 result, we (broadly speaking) see the point at which a theory moves from prediction to (broadly speaking) accepted fact, and in the Fraunhofer redshift observations we see the effect this has on a related field, turning relativistic effects into a background detail for solar spectroscopic measurements. The fact that Dyson and Eddington are so scrupulously exhaustive in describing their data analysis illustrates the continuing force of the Royal Society's motto 'Nullius in verba', conventionally glossed as 'take nobody's word for it' – the idea that scientific results should be described in a way that lets the reader repeat the observation, or re-do the analysis, themself, and which has logical, methodological and even ethical dimensions; the fact that no reader seems to have thought this re-observation worth doing illustrates the continuing force of scientists' communal trust in each other's craft skills, and in their professional honesty, in reporting and analysing data as fairly as possible.

So if, as I assert, this incident is more intricate than most people would expect, but not more intricate than usual practice, we must ask why it has become so famous, and so frequently and combatively discussed. Why, for that matter, is this usually described as Eddington's observation, though it seems to have been a joint effort, with Dyson, in every important respect? One answer is that it is a good story, in the practical sense that it incorporates the multiple features I mentioned above, in a way which illustrates the strands of motivation and persuasion which are present in actual scientific practice. Another answer is that it incorporates human involvement with the process – the result was personally important to Eddington, for more than merely professional reasons. I think it is those human aspects of the story – aspects which are essential to the story – which make some people uncomfortable, as if they suggest that scientific conclusions are arbitrary, or personal, or are somehow mere consensus (in a deprecatory sense, which is distinct from the pragmatic sense in which I have used the word above), and as if they undermine the thoughtful confidence we should have in scientific conclusions here and elsewhere. But that confidence is emphatically *not* undermined by the observation that scientists are people, that the logic of science is not only the simplistic one described in popular science books (or press releases), and that uncertainty and doubt are integral to the process.

So after 'a closer look' at the 1919 observations, what conclusions can we come to? The 'closer look' appears to have simply made things more complicated, and added specifically human complications to what seemed like a reassuringly rational, or logical, observation. A key thing to note is that we are asking at least *three* questions here, namely (i) is general relativity correct? (ii) how do astronomers conclude that general relativity is correct? and (iii) how did astronomers conclude that in the past? The last question is a historical and social question: what happened, as a matter of history, in the year-long run-up to the 1919 November 6 meeting where Dyson and Eddington reported their results, and what were the mechanisms through which the attendees at that meeting decided, as a matter of 'theory choice', that Dyson and Eddington had proved Einstein correct? Phrased like that, it is surely obvious that subsequent observational work – from later eclipse observations to the LIGO result – is simply irrelevant to the question. A related question is to ask how can astronomers *now* address that same 'theory choice'? They *can* use the hindsight of a century of further observation, and it is through that hindsight that this question shades into the first one: how do we know that relativity is correct? To answer that question would take us into the philosophy and pragmatics of science, and thus even further adrift from our main topic.

Although the history is complicated, we remain supremely confident (while respecting the detailed qualifications of Section B.2) that GR is a reliably correct description of nature.

Our appreciation of the history of the observations makes it all the more impressive that the evolution of scientific culture has produced a process which is able to accommodate the apparent methodological problems described on p. 188, working within a world of uncertainty, confidence, ambition, ambivalence, patriotism, bloody-mindedness, and every other human passion, which *still* ratchets towards a certain type of truth. Science does not remove these passions, but the process of accommodation is worthy of admiration, and worthy of examination and critical study, and possibly even of emulation. If we are clearer about science's boundaries, when we look at its work in current controversies, then we can be a little clearer about what is outside those boundaries – in particular topics of social or human policy where science or scientific approaches may have little to contribute – and a little clearer about the great confidence we can have in science's conclusions and assistance when it is inside those boundaries, asking the questions it can answer best.

# Appendix C

## Maths Revision

In a couple of places in the text I rely on a little familiarity with particular mathematical structures – namely complex numbers, hyperbolic 'trigonometry', and matrix algebra. I presume you have at least encountered these before, but appreciate that this text may be the first time you've used them outside of a maths class. I've therefore included, below, some very compact introductions to these bits of maths, which are self-contained, but probably more useful as a reminder than as a fully standalone introduction.

### C.1 Complex Numbers

We are most familiar with the *real numbers*, $\mathbb{R}$, which corresponds directly to the number line, and using which we can define addition, subtraction, multiplication, division, and so on. We might, however, ask ourselves, 'what about the number $x$, which is such that $x^2 = -1$? Where is that number on the number line?' One answer to this is that this simply isn't a number, or that the equation $x^2 = -1$ is meaningless.

Another answer is to say: what if there *is* a 'number' which is a solution to this equation – call it i – what are its properties? Can we define operations of addition, subtraction, multiplication and division with such numbers? You will not be surprised to discover that we can; you may be surprised to discover that this extra 'number' i is the *only* new thing we need, to uncover a completely new set of numbers. We refer to this set as the **complex numbers**, $\mathbb{C}$.

The rules for complex numbers follow fairly naturally:

**The number i is a 'complex number'** The number i is defined such that $i^2 = -1$, and is the prototype complex number.

**Any real multiple of i is a complex number** Thus 2i and $-999i$ and $\pi \times i$ are complex numbers. Multiples of i are known as *imaginary numbers* (there's nothing unreal about them, but this is the usual name, in contrast to the *real* numbers).

**We can add real numbers and complex numbers** So $1 + i$ is a complex number, as is $2 + i - 3 + i4 = -1 + i5$. Here, we have gathered the real numbers in this sum, and gathered the multiples of i, but we cannot simplify this expression further. That means that...

**All complex numbers can be expressed in the form $z = a + ib$** for $a, b \in \mathbb{R}$. That is, we can gather (and simplify) the terms involving i, and the terms not involving i, but we can't simplify further. We refer to $a$ as the *real part* of the number $z$, and $b$ as the *imaginary part*.

**All real numbers are complex numbers** That is to say, $\mathbb{R} \subset \mathbb{C}$, since the real numbers are just those complex numbers for which $b$, above, is zero.

The above points give us enough information to describe the full range of arithmetic operations.

Addition: Given two complex numbers $z_1 = a + ib$ and $z_2 = c + id$, the sum of the two numbers is obtained by collecting and simplifying as much as possible:

$$z_1 + z_2 = (a + ib) + (c + id) = (a + c) + i(b + d)$$
$$z_1 - z_2 = (a + ib) - (c + id) = (a - c) + i(b - d).$$

Multiplication: The product of $z_1$ and $z_2$ is obtained similarly, by multiplying the numbers out, and simplifying as much as possible:

$$z_1 z_2 = (a + ib)(c + id)$$
$$= ac + ibc + iad + i^2 bd$$
$$= (ac - bd) + i(bc + ad),$$

using the fact that $i^2 = -1$.

Division: We can define division slightly indirectly, relying on our definition of multiplication:

$$\frac{z_1}{z_2} = \frac{a + ib}{c + id}$$
$$= \frac{a + ib}{c + id} \times \frac{c - id}{c - id} \qquad \text{(multiplication by 1)}$$
$$= \frac{(ac + bd) + i(bc - ad)}{c^2 + d^2}.$$

Either using this, or by inspection, you can see that $1/i = -i$.

The complex conjugate: There is a new arithmetic operation we can perform on complex numbers, the *complex conjugate*: this is the operation of negating the imaginary part of the number. Thus, given a complex number $z = a + ib$, the conjugate, $z^*$, is

$$z^* = a - ib.$$

This is sometimes written $\bar{z}$. Using this, we can define the *modulus* of a complex number as $|z|$, where

$$|z|^2 = zz^* = (a + ib)(a - ib) = a^2 + b^2.$$

This corresponds to the pythagorean distance in the complex plane.

With a limited set of exceptions, all of the arithmetic operations we can use with real numbers, we can also use with complex numbers. The only reason I'm mentioning complex numbers in this context is because in Chapter 5 we use the hyperbolic functions, which are most naturally related to the similar trigonometric functions, but using complex numbers.

## C.2  The Hyperbolic Functions

You are, I presume, familiar with the trigonometric functions $\sin\theta$, $\cos\theta$, and so on. And so you will have seen the range of identities associated with those: $\sin^2\theta + \cos^2\theta = 1$, $\sec^2\theta = 1 + \tan^2\theta$, and the various addition formulae.

There is a related set of functions, called the hyperbolic functions (sometimes called the 'hyperbolic' or 'complex trigonometric functions', plausibly but inaccurately). These are defined by

$$\sinh z = \frac{1}{2}(e^z - e^{-z}) \tag{C.1a}$$

$$\cosh z = \frac{1}{2}(e^z + e^{-z}), \tag{C.1b}$$

and illustrated in Figure C.1. There are corresponding definitions

$$\tanh z = \frac{\sinh z}{\cosh z}$$

$$\operatorname{sech} z = \frac{1}{\cosh z},$$

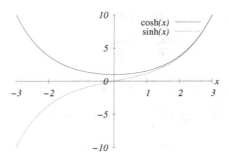

**Figure C.1** The sinh $z$ and cosh $z$ functions.

and identities such as

$$\cosh^2 x - \sinh^2 z = 1$$
$$1 - \tanh^2 z = \operatorname{sech}^2 z$$
$$\sinh(u \pm v) = \sinh u \cosh v \pm \cosh u \sinh v$$
$$\cosh(u \pm v) = \cosh u \cosh v \pm \sinh u \sinh v$$
$$\tanh(u \pm v) = \frac{\tanh u \pm \tanh v}{1 \pm \tanh u \tanh v}$$

See your favourite source of mathematical tables for the rest, and for expressions for the derivatives, and so on. You can see that these are quite similar to the corresponding trigonometric identities.

We can find other expressions for these functions by looking at the series expansion of the exponential function:

$$e^z = 1 + z + \frac{z^2}{2!} + \frac{z^3}{3!} + \frac{z^4}{4!} + \cdots. \tag{C.2}$$

By looking at Eq. (C.1), we can calculate that

$$\sinh z = z + \frac{z^3}{3!} + \frac{z^5}{5!} + \cdots$$
$$\cosh z = 1 + \frac{z^2}{2!} + \frac{z^4}{4!} + \cdots.$$

We may also now ask what is the exponential of a pure imaginary number, $iz$? This seems an odd thing to do, if you think of $e^z$ only as the number e raised to a given power; but if you think of Eq. (C.2) as *defining* the exponential function, then it is obvious what to do:

$$e^{iz} = 1 + iz + \frac{(iz)^2}{2!} + \frac{(iz)^3}{3!} + \frac{(iz)^4}{4!} + \cdots$$

$$= 1 + iz - \frac{z^2}{2!} - i\frac{z^3}{3!} + \frac{z^4}{4!} + \cdots$$

$$= \left(1 - \frac{z^2}{2!} + \frac{z^4}{4!} + \cdots\right) + i\left(z - \frac{z^3}{3!} + \frac{z^5}{5!} + \cdots\right)$$

$$= \cos z + i\sin z. \tag{C.3}$$

The final line comes when we spot the well-known series for $\cos z$ and $\sin z$,

$$\cos z = 1 - \frac{z^2}{2!} + \frac{z^4}{4!} + \cdots, \qquad \sin z = z - \frac{z^3}{3!} + \frac{z^5}{5!} + \cdots.$$

From Eq. (C.3), or directly from Eq. (C.2), we can now see that

$$\cos\theta = \frac{1}{2}(e^{i\theta} + e^{-i\theta}) \qquad\qquad \Rightarrow \cos i\theta = \cosh\theta$$

$$\sin\theta = \frac{1}{2i}(e^{i\theta} - e^{-i\theta}) \qquad\qquad \Rightarrow \sin i\theta = i\sinh\theta.$$

The argument in Section 5.1 used complex arguments to the trigonometric functions, in a way that may be surprising at first, but which turns out to be a straightforward application of the identities we see here.

Equation (C.3), incidentally, is known as *Euler's equation*, and rewritten in the special case of $z = \pi$, it is

$$e^{i\pi} + 1 = 0,$$

which Richard Feynmann called 'the most remarkable equation in mathematics,' involving as it does five of the key numbers in mathematics, in a single equation.

## C.3 Linear Algebra

The term 'linear algebra' refers to a quite general area of mathematics, but the way we most often, or first, encounter it as physicists is through the arithmetic of matrices. We encounter it first in this text in Section 6.1.

The key results are as follows.

If $\mathbf{a} = (a_x, a_y, a_z)$ and $\mathbf{b} = (b_x, b_y, b_z)$ are vectors, then $\mathbf{a} + \mathbf{b}$ is also a vector, with components $(a_x + b_x, a_y + b_y, a_z + b_z)$. We can define a *scalar product* $\mathbf{a} \cdot \mathbf{b} = a_x b_x + a_y b_y + a_z b_z$. We can define the *length*, $|\mathbf{a}|$, of a 3-vector in terms of the scalar product of a vector with itself: $|\mathbf{a}|^2 = \mathbf{a} \cdot \mathbf{a} =$

$a^2 = a_x^2 + a_y^2 + a_z^2$ (we can also see this from Pythagoras's theorem, or indeed from Eq. (6.1)), and we know that this is an invariant of a rotation – that is, that it takes the same value irrespective of the coordinate system. It is easy to see that $|\mathbf{a} + \mathbf{b}|^2 = a^2 + b^2 + 2\mathbf{a} \cdot \mathbf{b}$ and so, since both $a^2$ and $b^2$ are frame-invariant, the scalar product $\mathbf{a} \cdot \mathbf{b}$ must be frame-invariant also, even though the individual coordinates $a_i$ and $b_i$ are not. Finally, if the scalar product of two vectors vanishes, $\mathbf{a} \cdot \mathbf{b} = 0$, we say that the two vectors are *orthogonal*. If a euclidean vector is orthogonal to itself ($\mathbf{a} \cdot \mathbf{a} = 0$) then we can deduce that $a_i = 0$. In linear algebra, the scalar product is more generally termed the *inner product*, and the length of a vector is termed its *norm*.

In Minkowski space, in contrast, the scalar product of Section 6.2 can be negative (and a mathematician would therefore insist that it cannot, strictly, be called an 'inner product'), which means that, if the scalar product of two Minkowski vectors is zero, $\mathbf{A} \cdot \mathbf{A} = 0$, we therefore cannot deduce that $\mathbf{A} = 0$. Also, in this context, we can continue to talk of the magnitude of a vector, $\mathbf{A} \cdot \mathbf{A}$, but can no longer (and indeed need no longer) talk of the norm as the square root of this.

# Appendix D

## How to Do Calculations: a Recipe

### D.1 Key Things to Remember

Two frames are in standard configuration (see Section 1.7 and Figure 1.5) when:

1. they are aligned so that the $(x, y, z)$ and $(x', y', z')$ axes are parallel;
2. the frame $S'$ is moving along the $x$-axis with velocity $v$;
3. we set the zero of the time coordinates so that the origins coincide at $t = t' = 0$ (which means that the origin of the $S'$ frame is always at position $x = vt$).

All three conditions must be satisfied in order for the Lorentz transformation equations to be valid.

The values $x$, $t$, and so on, are *coordinates* of events in the unprimed frame, $S$ (they're not displacements, or lengths, or the intervals between events); the values $x'$, $t'$, and so on, are coordinates *of the same events*, as measured in $S'$.

Remember: an event is not 'in' one frame or another. An event is a frame-independent thing which is 'in' *every* frame. The difference between frames is that an event will have different *coordinates* in different frames.

### D.2 A Checklist for Relativity Problems

The process of solving a relativity problem is, in many cases, pretty mechanical. That's not the same as *easy*, but there are pretty clear steps you can go through to help you to the answers.

You won't necessarily go through each of these items in precisely this order – so this is perhaps more of a checklist than a recipe – but you'll almost

200

certainly need to complete all of the 'identify' steps before it makes sense to do either of the 'write down' steps.

**Draw the Minkowski diagram** Drawing the diagram is generally a useful step, since it forces you to work through the textual information in the question, turning it into events. It may make sense to draw the Minkowski diagram in both the 'stationary' and the 'moving' frame. But that requires...

**Identify the frames** Usually these will be pretty obvious – something is moving past something else – but always make it explicit.

**Identify the origins of the frames** Very often, one of the events will be at the origin, but sometimes it will be more natural to have something else there (for example, the centre of the Earth, or the centre of a rocket). Make sure you *write down* what's at the origins of the two frames, when, and make sure that your choice is compatible with standard configuration ($x = x' = 0$ and $t = t' = 0$). If you have *nothing* at the origin, think again: you're very probably making things hard for yourself.

**Identify the events** Relativity is all about the coordinates of events, as measured by observers in different frames. Remember events are things like explosions, or hand-claps, or light-flashes, that happen at a specific place, at an instant of time. If no event is described in the question, make one up (for example 'the train snaps a ribbon at the exit of the tunnel'). Very often, an event will be at the intersection of two worldlines.

**Identify the worldlines** Something will be moving! Identify what its worldline is in the 'stationary frame'. Identify the worldlines of anything *not* moving in this frame. Look at the description of each event: what worldline(s) is it on? Sometimes worldlines will join the dots between events; sometimes an event is, in effect, defined as happening at the place where two worldlines cross.

**Write down what you know** There will be numeric values in the question. What events do they correspond to? If it's a distance or a time for an event, it's a distance or time in which frame? Whenever you write down a space or time coordinate, you *must* make clear which event it is the coordinate of: if it is (for example) the time coordinate of event ② in the primed frame, write $t'_2$ (and not just $t'$).

**Write down what you have to find out** The question will probably (implicitly) require you to find the position or time coordinate of an event as viewed in the 'other' frame. What coordinate ($x_1$, say, or $t'_2$?)

are you being asked to find?

**Redraw your Minkowski diagram?** At this point you might have
changed your mind about your Minkowski diagram. That's OK: the
diagram is there to help you sort these steps out.

**Head for home** At this point, the rest is pretty mechanical. Use one of
the Lorentz transformation equations to find the coordinate you need
from the coordinates you have. You may have to do some rearrangement. There are a few sanity-checks you can do at this point. If two
events are connected by a light flash, do their coordinates show that
the light is moving at speed $c$ in all frames? Have lengths contracted,
and times dilated, as you might expect? These checks are the new
intuitions you acquire through doing SR problems.

The first steps are the hard ones. After about half-way, you're just turning
the handle.

The big wrinkle is that *you might not need to use the Lorentz transformation*, since there are a few other bits of dynamics you're supposed to
know.

## D.3  Which Equation When?

**Lorentz transformation equations** These relate the coordinates of
*one* event in *two* frames. Remember that, for the Lorentz transformation equations to work, the two frames *must* be in standard
configuration. Given, say, $x_1$ and $t_1$, you might be asked to find $x_1'$
and/or $t_1'$ (all coordinates of the *same* event ①). **Whenever you use
one of the Lorentz transformation equations, you *must* identify
which event you're talking about.** If you write down a Lorentz
transformation equation as part of a calculation, and the $x$s and $t$s
don't have subscripts identifying an event (the *same* event), *then you're
doing it wrong*.

**Length contraction and time dilation** These are for calculating *intervals* – the length of a stick, or the interval between clock ticks – in
one frame, as measured in another. You can deduce them from the
Lorentz transformation, but also (as we saw in Chapter 3) directly from
the axioms. You won't necessarily have to set up frames in standard
configuration in order to use these relations.

**Distance is speed times time** This is for *movement within a frame*
(possibly including the movement of one frame with respect to an-

other). If something is moving at speed $v$ in some frame, for a time $\Delta t$, then it will have moved a distance $v\Delta t$ *in that frame*.

## D.4 Rest, Moving and Stationary Frames: Be Careful!

Be careful when using the terms 'rest frame' or 'moving frame'.

Frame $S'$ will often be referred to as the *rest frame*; however, it should *always* be the rest frame *of something*. Yes, it does seem a little counter-intuitive that it's the 'moving frame' that's the rest frame, but it's called the rest frame because it's the frame in which the thing we're interested in – be it a train carriage or an electron – is at rest. It's in the rest frame of the carriage that the carriage is measured to have its *rest length* or *proper length*. In general, talking of the 'rest', 'moving' and 'stationary' frames is a bit sloppy, because (and this is part of the point) what's 'moving' and what's 'stationary' is relative to who's doing the observing: the carriage is moving in the observer's rest frame, and the observer is moving in the carriage's rest frame. For this reason it's generally better to talk of 'the frame of the station platform' (or equivalently 'the rest frame of the platform') or 'the (rest) frame of the particle', but to avoid sounding pointlessly fussy, we'll stick to the more informal version most of the time, if the identity of the frames is clear from context. But whenever you see or write 'rest frame' or 'stationary frame', you should *always* ask the question '*whose* frame?'.

Similarly, it never makes sense to talk of 'the primed frame' or 'the un-primed frame' to refer to 'the moving frame' *unless* you've previously made clear which frame is which, through a diagram or through a textual explanation. Usually, for the sake of consistency, we draw diagrams so that the 'moving' frame is the primed one, but that's just a neat convention, and not anything you can start a calculation with.

You should be extremely precise here, not just for the sake of your reader (though remember that that reader might be your exam marker!), but because being fussily precise here helps you organise a problem in your head, and so make a good start on solving it. Quite a lot of relativity problems are actually quite simple *if* (and **only if**) people are clear in their own head what they're talking about; and when people tie themselves in knots (for example in exam answers), it is often because they have failed to set things out in a systematic way (such as Section D.2).

## D.5  Length Contraction, Time Dilation, and 'Rest Frames'

$L_0$ and $T_0$ are the length and proper time in the *rest frame* of a particle/object/train. That may or may not be the 'primed frame'.

For length contraction and time dilation, I don't try to memorise an equation such as Eq. (3.3) or Eq. (3.5): instead, I simply remember that 'length contraction' means that 'moving rods get shorter', and 'time dilation' means that 'moving clocks run slow'. Make sure you understand what these slogans mean, and multiply or divide by $\gamma$ as appropriate.

Finally, be careful about terminology. *The rest frame of a particle (or train, or person) is the frame in which that particle is not moving* (everyone agrees about this frame). The 'moving frame' and 'stationary frame' are informal and relative terms – the station is the 'moving frame' to an observer on the train. Similarly, the 'primed frame' is usually chosen as the 'moving frame', but this is entirely arbitrary.

# References

Adelberger, E., Gundlach, J., Heckel, B. et al. (2009), 'Torsion balance experiments: A low-energy frontier of particle physics', *Progress in Particle and Nuclear Physics* **62**(1), 102–134. https://doi.org/10.1016/j.ppnp.2008.08.002.

Adlam, E. (2011), Poincaré and special relativity. Preprint, https://arxiv.org/abs/1112.3175.

Ashby, N. (2003), 'Relativity in the global positioning system', *Living Reviews in Relativity* **6**(1), 1. https://doi.org/10.12942/lrr-2003-1.

Bailey, J., Borer, K., Combley, F. et al. (1977), 'Measurements of relativistic time dilatation for positive and negative muons in a circular orbit', *Nature* **268**(5618), 301–305. https://doi.org/10.1038/268301a0.

Barton, G. (1999), *Introduction to the Relativity Principle*, John Wiley and Sons. ISBN 9780471998969.

Bell, J. S. (1976), 'How to teach special relativity', *Progress in Scientific Culture* **1**(2), 135–148. Reprinted with minor changes in Bell (2004, ch. 9).

Bell, J. S. (2004), *Speakable and Unspeakable in Quantum Mechanics: Collected Papers on Quantum Philosophy*, 2nd ed., Cambridge University Press. ISBN 9780521523387.

BIPM (2019), 'Le système international d'unités / the international system of units ('the SI brochure')', Online. https://www.bipm.org/en/publications/si-brochure.

Bloor, D. (1981), 'II.2 The strengths of the strong programme', *Philosophy of the Social Sciences* **11**(2), 199–213. https://doi.org/10.1177/004839318101100206.

Bloor, D. (1991), *Knowledge and Social Imagery*, 2nd ed., University of Chicago Press. First edition 1976, ISBN 9780226060972.

Collins, H. M. & Pinch, T. (2012), *The Golem: What Everyone Should*

*Know about Science*, 2nd ed., Cambridge University Press. ISBN 9781107604650.

Dewan, E. M. (1963), 'Stress effects due to Lorentz contraction', *American Journal of Physics* **31**(5), 383. https://doi.org/10.1119/1.1969514.

Dewan, E. M. & Beran, M. (1959), 'Note on stress effects due to relativistic contraction', *American Journal of Physics* **27**(7), 517–518. https://doi.org/10.1119/1.1996214.

Dyson, F. W., Eddington, A. S. & Davidson, C. (1920), 'IX. A determination of the deflection of light by the sun's gravitational field, from observations made at the total eclipse of May 29, 1919', *Philosophical Transactions of the Royal Society of London. Series A* **220**(579), 291–333. https://doi.org/10.1098/rsta.1920.0009.

Earman, J. & Glymour, C. (1980*a*), 'The gravitational red shift as a test of general relativity: History and analysis', *Studies in History and Philosophy of Science Part A* **11**(3), 175–214. https://doi.org/10.1016/0039-3681(80)90025-4.

Earman, J. & Glymour, C. (1980*b*), 'Relativity and eclipses: The British eclipse expeditions of 1919 and their predecessors', *Historical Studies in the Physical Sciences* **11**(1), 49–85. https://doi.org/10.2307/27757471.

Eddington, A. S. (1917), 'Karl Schwarzschild', *Monthly Notices of the Royal Astronomical Society* **77**(4), 314–319. https://doi.org/10.1093/mnras/77.4.314.

Eddington, A. S. (1920), *Report on the Relativity Theory of Gravitation*, 2nd ed., Physical Society of London. Revised edition of the original 1918 publication, https://archive.org/details/reportontherelat028829mbp.

Eddington, A. S. (1940), 'Sir Frank Watson Dyson, 1868–1939', *Obituary Notices of Fellows of the Royal Society* **3**(8), 159–172. https://doi.org/10.1098/rsbm.1940.0015.

Einstein, A. (1905), 'Zur Elektrodynamik bewegter Körper (on the electrodynamics of moving bodies)', *Annalen der Physik* **17**, 891. Reprinted in translation in Lorentz et al. (1952), https://doi.org/10.1002/andp.19053221004.

Einstein, A. (1911), 'Über den Einfluss der Schwerkraft auf die Ausbreitung des Lichtes (on the influence of gravitation on the propagation of light)', *Annalen der Physik* **35**, 898–908. Reprinted in translation in Lorentz et al. (1952), https://doi.org/10.1002/andp.19113401005.

Einstein, A. (1916), 'Die Grundlage der allgemeinen Relativitätstheorie (the foundation of the general theory of relativity)', *Annalen der Physik* **49**(7), 769–822. Reprinted in Lorentz et al. (1952); additionally online at https://einsteinpapers.press.princeton.edu/vol6-trans/158, https://

doi.org/10.1002/andp.19163540702.

Einstein, A. (1920), *Relativity: The Special and the General Theory*, Methuen. Originally published in book form in German, in 1917; first published in English in 1920, in an authorised translation by Robert W. Lawson; available in multiple editions and formats.

Einstein, A. (1936*a*), 'Physics and reality', *Journal of the Franklin Institute* **221**(3), 349–382. Translation, by Jean Piccard, of Einstein (1936*b*), https://doi.org/10.1016/S0016-0032(36)91047-5.

Einstein, A. (1936*b*), 'Physik und Realität', *Journal of the Franklin Institute* **221**(3), 313–347. Translated in Einstein (1936*a*), https://doi.org/10.1016/S0016-0032(36)91045-1.

Einstein, A. (1991), *Autobiographical Notes*, Open Court. First published in a separate edition 1979; various printings, ISBN 9780812691795.

Everitt, C., Lipa, J. & Siddall, G. (1979), 'Precision engineering and Einstein: The relativity gyroscope experiment', *Precision Engineering* **1**(1), 5–11. https://doi.org/10.1016/0141-6359(79)90070-9.

FitzGerald, G. F. (1889), 'The ether and the Earth's atmosphere', *Science* **13**(328), 390. https://doi.org/10.1126/science.ns-13.328.390.

French, A. P. (1968), *Special Relativity*, CRC Press. ISBN 9780748764228.

Galileo Galilei (1632), *Dialogue Concerning the Two Chief World Systems*, Batista Landini. Original title (in effect): *Dialogo sopra i due massimi sistemi del mondo*.

Gourgoulhon, E. (2013), *Special Relativity in General Frames: From Particles to Astrophysics*, Springer-Verlag. ISBN 9783662520833.

Gray, N. (2019), *A Student's Guide to General Relativity*, Cambridge University Press. ISBN 9781107183469.

Hamill, P. (2013), *A Student's Guide to Lagrangians and Hamiltonians*, Cambridge University Press. ISBN 9781107617520.

Harvey, G. M. (1979), 'Gravitational deflection of light', *Observatory* **99**, 195–198. https://ui.adsabs.harvard.edu/abs/1979Obs....99..195H.

Hawking, S. W. (1988), *A Brief History of Time*, Bantam.

Heaviside, O. (1889), 'On the electromagnetic effects due to the motion of electrification through a dielectric', *Philosophical Magazine (fifth series)* **27**(167), 324–339. https://doi.org/10.1080/14786448908628362.

Hentschel, K. (1996), 'Measurements of gravitational redshift between 1959 and 1971', *Annals of Science* **53**(3), 269–295. https://doi.org/10.1080/00033799600200211.

Jefimenko, O. D. (1994), 'Direct calculation of the electric and magnetic fields of an electric point charge moving with constant velocity', *American Journal of Physics* **62**(1), 79–85. https://doi.org/10.1119/1.17716.

Kennefick, D. (2009), 'Testing relativity from the 1919 eclipse—a question of bias', *Physics Today* **62**(3), 37–42. https://doi.org/10.1063/1.3099578.

Kennefick, D. (2012), 'Not only because of theory: Dyson, Eddington, and the competing myths of the 1919 eclipse expedition', in C. Lehner, J. Renn & M. Schemmel, eds, *Einstein and the Changing Worldviews of Physics*, Birkhäuser, Boston, pp. 201–232. https://doi.org/10.1007/ 978-0-8176-4940-1_9.

Kuhn, T. S. (1996), *The Structure of Scientific Revolutions*, 3rd ed., Chicago University Press. ISBN 978-0226458083.

Labinger, J. A. & Collins, H., eds (2001), *The One Culture?: A Conversation About Science*, Chicago University Press. ISBN 9780226467238.

Landau, L. D. & Lifshitz, E. M. (1975), *The Classical Theory of Fields*, Vol. 2 of *Course of Theoretical Physics*, 4th ed., Butterworth-Heinemann. ISBN 9780750627689.

Laudan, L. (1981), 'II.1 The pseudo-science of science?', *Philosophy of the Social Sciences* **11**(2), 173–198. See also Bloor (1981), https://doi.org/10. 1177/004839318101100205.

Longair, M. S. (2020), *Theoretical Concepts in Physics: An Alternative View of Theoretical Reasoning in Physics*, 3rd ed., Cambridge University Press. ISBN 9781108484534.

Lorentz, H. A. (1895), 'Michelson's interference experiment'. Reprinted in Lorentz et al. (1952), as a translation of §§89–92 of Lorentz's book *Versuch einer Theorie der elektrischen und optischen Erscheinungen in bewegten Körpern*, Leiden, 1985.

Lorentz, H. A. (1904), 'Electromagnetic phenomena in a system moving with any velocity less than that of light', *Proceedings of the Academy of Sciences of Amsterdam* **6**, 809–831. Reprinted in Lorentz et al. (1952), https://ui.adsabs.harvard.edu/abs/1903KNAB....6..809L.

Lorentz, H. A., Einstein, A., Minkowski, H. et al. (1952), *The Principle of Relativity*, Dover. ISBN 9780486600819.

Mattingly, D. (2005), 'Modern tests of Lorentz invariance', *Living Reviews in Relativity* **8**(5). https://doi.org/10.12942/lrr-2005-5.

Mermin, N. D. (1990), *Boojums All the Way Through*, Cambridge University Press. ISBN 9780521388801.

Miller, D. C. (1933), 'The ether-drift experiment and the determination of the absolute motion of the earth', *Reviews of Modern Physics* **5**, 203–242. https://doi.org/10.1103/RevModPhys.5.203.

Minkowski, H. (1908), 'Space and time', An Address delivered at the 80th Assembly of German Natural Scientists and Physicians, Cologne. Reprinted in Lorentz et al. (1952).

Misner, C. W., Thorne, K. S. & Wheeler, J. A. (1973), *Gravitation*, Freeman. ISBN 0-7167-0334-3.

Pais, A. (2005), *'Subtle is the Lord ...': The Science and the Life of Albert Einstein*, revised ed., Oxford University Press. ISBN 9780192806727.

Pascoli, G. (2017), 'The Sagnac effect and its interpretation by Paul Langevin', *Comptes Rendus Physique* **18**(9), 563–569. https://doi.org/10.1016/j.crhy.2017.10.010.

Reif, F. (1961), 'The competitive world of the pure scientist', *Science* **134**(3494), 1957–1962. https://doi.org/10.1126/science.134.3494.1957.

Rindler, W. (1961), 'Length contraction paradox', *American Journal of Physics* **29**(6), 365–366. https://doi.org/10.1119/1.1937789.

Rindler, W. (1967), 'World's fastest way to get the relativistic time-dilation formula', *American Journal of Physics* **35**(12), 1165–1165. https://doi.org/10.1119/1.1973819.

Rindler, W. (1977), *Essential Relativity: Special, General and Cosmological*, 2nd ed., Springer-Verlag. ISBN 9783540100904.

Rindler, W. (2006), *Relativity: Special, General, and Cosmological*, 2nd ed., Oxford University Press. ISBN 9780198567325.

Roberts, T. & Schleif, S. (2007), 'What is the experimental basis of special relativity?', Web page. http://math.ucr.edu/home/baez/physics/Relativity/SR/experiments.html.

Schutz, B. F. (2009), *A First Course in General Relativity*, 2nd ed., Cambridge University Press. ISBN 9780521887052.

Schwartz, J. & McGuinness, M. (2012), *Introducing Einstein*, Icon. First published in 1979; this book clearly has a messy publication history: multiple successive publishers seem to have alternated its name between this and 'Einstein for Beginners', ISBN 9781848314085.

Schwarz, P. M. & Schwarz, J. H. (2004), *Special Relativity: From Einstein to Strings*, Cambridge University Press. ISBN 9780521812603.

Shankland, R. S., McCuskey, S. W., Leone, F. C. et al. (1955), 'New analysis of the interferometer observations of Dayton C. Miller', *Reviews of Modern Physics* **27**, 167–178. https://doi.org/10.1103/RevModPhys.27.167.

Takeuchi, T. (2012), *An Illustrated Guide to Relativity*, Cambridge University Press. ISBN 9780511779121.

Taylor, E. F. & Wheeler, J. A. (1992), *Spacetime Physics*, 2nd ed., W. H. Freeman. See also https://www.eftaylor.com/spacetimephysics/, ISBN 9780716723271.

Taylor, E. F., Wheeler, J. A. & Bertschinger, E. (2019), *Exploring Black Holes*, 2nd ed., online. Public draft as of 2021, https://www.eftaylor.com/exploringblackholes/.

Thomson, J. J. (1889), 'On the magnetic effects produced by motion in the electric field', *Philosophical Magazine (fifth series)* **28**(170), 1–14. https://doi.org/10.1080/14786448908619821.

Thomson, J. J. (1919), 'Joint eclipse meeting of the Royal Society and the Royal Astronomical Society', *The Observatory* **42**, 389–398. https://ui.adsabs.harvard.edu/abs/1919Obs....42..389.

Vessot, R. F. C., Levine, M. W., Mattison, E. M. et al. (1980), 'Test of relativistic gravitation with a space-borne hydrogen maser', *Phys. Rev. Lett.* **45**, 2081–2084. https://doi.org/10.1103/PhysRevLett.45.2081.

Will, C. M. (2014), 'The confrontation between general relativity and experiment', *Living Reviews in Relativity* **17**(1), 4. https://doi.org/10.12942/lrr-2014-4.

# Index

Printed in the United States
by Baker & Taylor Publisher Services